CORROSION OF STAINLESS STEELS

THE CORROSION MONOGRAPH SERIES

R. T. Foley, N. Hackerman, C. V. King, F. L. LaQue, and Z. A. Foroulis, Editors

Friend, *Corrosion of Nickel and Nickel Base Alloys*
Sedriks, *Corrosion of Stainless Steels*
Frankenthal and Kruger, *Passivity of Metals*
LaQue, *Marine Corrosion*
Berry, *Corrosion in Nuclear Applications*
Ailor, *Handbook on Corrosion Testing and Evaluation*
Kofstad, *High-Temperature Oxidation of Metals*
Logan, *The Stress Corrosion of Metals*
Godard, Jepson, Bothwell, and Kane, *The Corrosion of Light Metals*
Leidheiser, *The Corrosion of Copper, Tin, and Their Alloys*

In Preparation (*Titles Tentative*)

Foroulis, *Aqueous Corrosion and Its Inhibition*
Ailor, *Atmospheric Corrosion of Metals*
Fehlner and Kofstad, *Oxidation of Metals*
Hackerman, McCafferty, and Snavely, *Inhibitors and Inhibition*

CORROSION OF STAINLESS STEELS

A. JOHN SEDRIKS

Corrosion Section Manager
INCO Research and Development Center
Sterling Forest
Suffern, New York

Sponsored by
THE ELECTROCHEMICAL SOCIETY, INC. *Princeton, New Jersey*

A Wiley-Interscience Publication
John Wiley & Sons *New York Chichester Brisbane Toronto Singapore*

Copyright © 1979 by John Wiley & Sons, Inc.

All rights reserved. Published simultaneously in Canada.

Reproduction or translation of any part of this work
beyond that permitted by Sections 107 or 108 of the
1976 United States Copyright Act without the permission
of the copyright owner is unlawful. Requests for
permission or further information should be addressed to
the Permissions Department, John Wiley & Sons, Inc.

Library of Congress Cataloging in Publication Data

Sedriks, A. John.
Corrosion of stainless steels.

(The Corrosion monograph series)
"Sponsored by the Electrochemical Society, Inc.,
Princeton, New Jersey."
"A Wiley-Interscience publication."
Includes index.
1. Steel, Stainless—Corrosion. I. Electrochemical
Society. II. Title. III. Series.

TA479.S7S4 620.1′7′23 79-11985
ISBN 0-471-05011-3

Printed in the United States of America

10 9 8 7 6 5 4

TA
479
S7
S4

To Sally, Andrew, Amanda and Sarah

PREFACE TO THE SERIES

The *Corrosion Monograph Series* is the mechanism chosen by the Corrosion Division of The Electrochemical Society to bring the corrosion literature up to date.

Since 1945, the field of corrosion has seen a tremendous amount of research and development stimulated by problems related to great advances in technology and facilitated by parallel advances in corrosion science. This has resulted in a corresponding expansion in the published literature. Indeed, in the late forties whole subdivisions of corrosion knowledge, such as that related to corrosion of titanium and to corrosion by ultra pure water at high temperature, were practically nonexistent.

Although knowledge in some areas of corrosion has continued to evolve at a less accelerated pace, data published in the mid-forties have been supplemented and supported by results of continuing studies that have provided an enlarged base for dealing with specific problems.

As a result of the appearance of numerous new concepts and such a large body of new data in so many areas, the total mass of information has become more than a single-volume treatise can contain. The preparation of monographs in selected areas, to be followed by appropriate revision, was chosen as the preferred course to keep pace with hoped-for progress.

These monographs are being written by specialists in each field, and the number of authors contributing to a single volume has been intentionally kept small to provide for such sharpening of point of view as is offered by small areas, but not large ones. Also, this accelerates publication.

The nature of the presentation, that is, the relative emphasis on the scientific aspects as contrasted to the engineering aspects, varies with the subject being dealt with, the nature of pertinent data, and the response of the authors to these influences. The several volumes include critical exposition of modern theories as well as specific data of the handbook type so useful to the practicing engineer. With this approach, the Series is designed to be useful to basic scientists working in research as well as to engineers confronted with corrosion problems in the field.

The Corrosion Monograph Series Committee

R. T. Foley, Chairman
N. Hackerman
C. V. King
F. L. LaQue
Z. A. Foroulis

PREFACE

This book is an introduction to the extensive subject of corrosion of stainless steels. It has been written mainly for engineers and students striving to acquire some background and perspective in this subject. Particular emphasis has been placed on describing how metallurgical factors affect the corrosion resistance of stainless steels. Information regarding their usage, corrosion problems, and their development and production has been drawn primarily from U. S. sources.

In reading this book, it should be borne in mind that stainless steels are widely used as materials of construction because of their corrosion resistance, and in many applications problems due to corrosion do not arise. However, attempts to extend their use to more demanding technological applications has continued to generate interest in their corrosion resistance, often with a view to defining limits for their use. In order to provide some perspective about materials used beyond such limits, it has also been necessary to include in this book some discussion about higher alloys.

Many of the tables and figures are used in this book by permission of copyright owners. I wish to acknowledge permission granted by American Chemical Society, Metals Society, McGraw-Hill Book Co., Pergamon Press, American Society for Testing and Materials, Welding Research Council, American Society for Metals, American Institute of Metallurgical Engineers, Newnes-Butterworths, Firth Brown Ltd., Cabot Corporation, Verlag Chemie GmbH, Her Majesty's Stationery Office, Institution of Chemical Engineers, Technische Ueberwachung, Brown Boveri, Institution of Civil Engineers, Sandvik Aktiebolag, Marshall Space Flight Center, Plenum Press, American Society of Mechanical Engineers, Naval Research Laboratory, Electrochemical Society, E. I.

duPont de Nemours Co., Sawell Publications, National Association of Corrosion Engineers, Foundry Trade Journal, Reinhold Publishing Co., Sumitomo Metals, Fuel and Metallurgical Journals Ltd., Edward Arnold (Publishers) Ltd., Morgan-Grampian Ltd., American Iron and Steel Institute, Japan Society of Corrosion Engineering, and John Wiley & Sons, Inc.

I would also like to express my gratitude to my many expert friends who helped me in the preparation of this book by providing key information or reviewing the text. Among many, I would especially like to thank Earle Hoxie, Tony Graae, and Norman Flint of Inco, Mike Henthorne of Cartech, and John Truman of Firth Brown, none of whom, however, bear any responsibility for any deficiencies. I would also like to thank Jerry Kruger of NBS and Ray Decker of Inco for their help and encouragement, and Marie Stachelski for her perseverance and accuracy in typing the manuscript.

A. JOHN SEDRIKS

Suffern, New York
April 1979

CONTENTS

CORROSION OF STAINLESS STEELS

1

INTRODUCTION

1-1 TYPES OF STAINLESS STEELS

Stainless steels represent less than 2% of the total amount of steel produced in the United States. However, because they are construction materials for key corrosion resistant equipment in most of the major industries, particularly in the chemical, petroleum, process, and power industries, they have a technological and economic importance far greater than would be indicated by the above percentage.

Stainless steels are iron alloys containing a minimum of approximately 11% chromium. This amount of chromium prevents the formation of rust in unpolluted atmospheres, as shown in Figure 1-1 (1); it is from this characteristic that their popular designation "stainless" is derived. Their corrosion resistance is provided by a very thin surface film which is self-healing in a wide variety of environments.

Today there are more than 170 different kinds of alloys that can be recognized as belonging to the stainless steel group, and each year new ones and modifications of existing ones appear. In some steels chromium content now approaches 30% and many other elements are added to provide specific properties or ease of fabrication. For example, nickel and molybdenum are added for corrosion resistance, carbon, molybdenum, titanium, aluminum, and copper for strength, sulfur and selenium for machinability, and nickel for formability and toughness.

It is customary to divide the more common stainless steels into three groups according to metallurgical structure: austenitic (face centered cubic), ferritic (body centered cubic), and martensitic (body centered tetragonal or cubic).

A simple, although somewhat approximate, way to relate metallurgical structure with the composition of stainless steels is by means of the Schaeffler diagram (2) modified by Schneider (3), which is shown in Figure 1-2. This diagram indicates the structure obtained after rapid

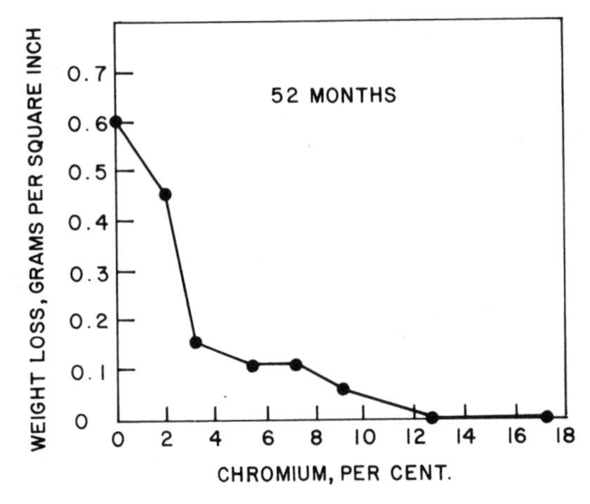

Figure 1-1. The influence of chromium on the atmospheric corrosion of low carbon steel. (After Binder and Brown.)

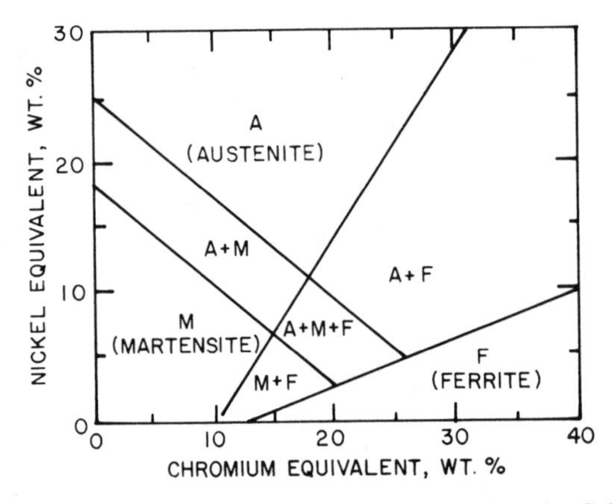

Figure 1-2. Schaeffler diagram including modifications by Schneider.

cooling to room temperature from 1050°C and is not an equilibrium diagram. It was originally established to estimate the delta ferrite content of welds in austenitic steels (2, 4).* In establishing this diagram, the alloying elements commonly found in stainless steels are regarded either as austenite stabilizers or delta ferrite stabilizers. The relative "potency" of each element is conveniently expressed in terms of either equivalence to nickel

*Delta ferrite is ferrite formed on solidification, as opposed to alpha ferrite which is the transformation product of austenite or martensite.

(austenite stabilizer) or chromium (ferrite stabilizer) on a weight percentage basis. The nickel and chromium equivalents, which form the two axes of the Schaeffler diagram, can be calculated from the equations given below (5):

$$\%\ \text{Ni equivalent} = \%\ \text{Ni} + \%\ \text{Co} + 30\ (\%\ \text{C}) + 25\ (\%\ \text{N}) + 0.5\ (\%\ \text{Mn}) + 0.3\ (\%\ \text{Cu})$$

$$\%\ \text{Cr equivalent} = \%\ \text{Cr} + 2\ (\%\ \text{Si}) + 1.5\ (\%\ \text{Mo}) + 5\ (\%\ \text{V}) + 5.5\ (\%\ \text{Al}) + 1.75\ (\%\ \text{Cb}) + 1.5\ (\%\ \text{Ti}) + 0.75\ (\%\ \text{W})$$

1-2 IDENTIFICATION OF STAINLESS STEELS

The American Iron and Steel Institute (AISI) designates the wrought standard grades of stainless steels by three digit numbers. The austenitic grades are designated by numbers in the 200 and 300 series, whereas the ferritic and martensitic grades are designated by numbers in the 400 series. For example, some of the more common austenitic steels are identified as types 201, 301, 304, 316, and 310, ferritic as types 430 and 446, and martensitic as types 410, 420, and 440C.* Duplex (austenitic-ferritic) stainless steels, precipitation hardening stainless steels, and higher alloys containing less than 50% iron are generally known by proprietary designations or trademarks and are referred to in this way throughout this book.† Ownership of trademarks is listed in the Appendix. Cast stainless steels are generally known by Alloy Casting Institute (ACI) designations which are described in Chapter 2.

When it comes to corrosion resistance in relatively severe environments, it is generally accepted that the austenitic stainless steels are superior. In milder environments ferritic stainless steels may have adequate corrosion resistance, whereas the martensitic and precipitation hardening stainless steels are considered useful in mildly corrosive environments where high strength or hardness is required. To complete the picture, it should be recognized that even the more highly alloyed austenitic stainless steels may not withstand some of the very severe environments which cause localized attack, such as pitting and crevice corrosion. In such instances search for resistant materials may be extended into the austenitic Ni-Cr-Fe-(Mo, Cu, Cb) alloys. Similarly, for high temperature water, caustic, or gaseous chlorine service, Ni-Cr-Fe alloys may

*A new alloy identification system, the Unified Numbering System (UNS), is being developed to describe all grades of stainless steel.

†The use of trademarks or other proprietary alloy designations has been necessary for purposes of exact description and is common practice. In no case does identification by trademarks imply recommendation or endorsement by any of the parties associated with this publication, nor does it imply that any material so identified is either the only material available or necessarily the best for any given application.

be considered. However, stainless steels provide acceptable corrosion resistance in numerous technological applications. In order to obtain a perspective as to which stainless steels are used to what extent, it is instructive to examine recent production figures and developments in steelmaking technology.

1-3 U. S. PRODUCTION OF STAINLESS STEELS

The relative ingot production of the various grades is often taken as a rough indicator of relative U. S. usage. The more recent ingot production figures reported by AISI available to date are shown in Table 1-1. Even a

TABLE 1-1. U.S. PRODUCTION OF STAINLESS STEEL INGOTS. PRODUCTION BY TYPE NUMBERS AS REPORTED BY AMERICAN IRON AND STEEL INSTITUTE (NET TONS)

Number	1971	1972	1973	1974	1975	1976	1977
201,202	65,267	56,125	49,631	39,276	30,797	46,082	47,525
301	99,088	149,556	193,038	152,276	96,979	193,647	182,719
302,302B	20,831	28,688	32,443	29,895	19,910	22,032	21,818
303	19,596	30,979	41,011	47,295	14,833	28,540	35,464
303Se	2,488	1,663	2,150	2,347	1,319	1,062	1,595
304	276,121	494,847	648,755	762,215	306,150	553,124	616,874
304L	37,848	43,616	53,361	67,942	51,107	47,247	57,401
305	9,677	9,907	11,480	12,555	4,562	9,736	8,383
308	2,772	2,090	4,647	5,696	1,805	2,095	2,632
309	6,215	5,711	7,979	9,468	5,669	4,255	4,022
309S	2,962	3,975	5,234	4,611	4,327	5,350	6,393
310	4,657	4,765	4,984	7,973	3,094	3,556	5,879
310S	471	792	728	--	746	1,504	2,568
316	53,773	63,476	102,175	129,079	58,057	72,100	87,752
316L	20,603	28,035	46,467	64,296	44,146	46,390	55,998
317	1,357	756	1,237	1,993	2,579	2,213	1,981
321	14,866	19,669	24,835	30,993	15,431	12,877	19,251
347	4,088	5,299	7,657	8,558	7,292	4,450	5,094
348	682	565	1,984	808	--	--	--
Other	58,333	80,197	99,362	101,002	65,501	78,075	87,646
Total	801,695	1,030,721	1,339,158	1,478,278	734,304	1,134,335	1,250,995
403	19,814	50,392	30,374	21,698	14,653	7,242	10,998
405,406	3,243	1,269	3,322	4,218	3,670	1,087	1,369
410	45,839	59,999	74,840	136,770	56,827	43,505	41,354
414	773	360	609	754	344	521	526
416,416Se	26,422	34,778	38,352	49,406	23,102	29,866	39,131
420	4,150	5,721	6,021	6,568	3,551	3,573	4,199
430	137,946	140,256	164,818	138,821	44,809	91,238	94,446
430F,430F Se	4,561	5,134	5,533	7,244	15,788	3,210	4,478
431	3,955	5,204	2,917	1,036	1,677	1,358	1,236
440A,440B	1,536	1,429	2,963	2,336	3,943	734	2,556
440C	6,061	8,122	10,303	9,965	5,933	6,422	7,806
442,443	807	723	1,214	1,184	423	796	260
446	1,707	2,570	2,174	1,725	1,385	1,159	1,477
Other	121,749	131,584	118,331	196,961	130,715	275,973*	360,494*
Total	378,518	447,541	461,771	581,784	306,820	466,684	570,330

*These figures reflect increased production of type 409 for automobile emission control equipment.

cursory inspection of these figures reveals the predominant position of the more corrosion resistant austenitic grades (300 series) over the ferritic and martensitic grades (400 series). By far the most popular grade in the 1970s has been type 304 austenitic stainless steel, which in most years has been produced in quantities exceeding those of the total production of the ferritic and martensitic grades. The production of type 316 has also shown an increase.

From a corrosion resistance viewpoint, of particular interest is the sustained production, even in the recession year 1975, of the extra low carbon grades, type 304L and type 316L. In these grades the carbon content is reduced to 0.03% (maximum) to improve resistance to sensitization which can cause corrosive attack at the grain boundaries after welding or high temperature exposure.

1-4 STAINLESS STEELMAKING PRACTICE

The ability to readily control the carbon content of stainless steels has been one factor, although by no means the only one, behind certain changes that are currently taking place in the production of stainless steels. The lowest carbon levels obtainable by various steelmaking processes are shown in Table 1-2. While a number of production techniques have been and are being used for stainless steels, a popular procedure has been to air melt the charge in an electric furnace and to refine by certain slag practices. Carbon content has been usually lowered by blowing with oxygen. However, this oxygen blowing is generally effective in reducing the carbon content to only about 0.04% before the onset of a serious loss of chromium by oxidation. To meet the extra low carbon specification of 0.03% (maximum), as in types 304L and 316L, it has been necessary in the past to use high purity (low carbon) charge materials. The use of low carbon charge materials also reduces the amount of chromium required in the charge, and has therefore been a popular practice.

TABLE 1-2. CARBON CONTENTS ATTAINABLE BY VARIOUS STEELMAKING PROCESSES

Steelmaking Process	% Carbon	
	Lowest	Typical
Basic Electric Arc	0.03[a]	0.05
Argon-Oxygen (AOD)	0.01-0.02	0.05
Electron Beam	--	0.001

[a]Attained by the use of low carbon charge materials.

In 1968 Union Carbide began commercializing a production technique known as Argon-Oxygen Decarburization (AOD), and this practice has met with considerable acceptance by U. S. stainless steel producers. In AOD practice carbon content is lowered by blowing with an argon-oxygen mixture. One of the advantages of this procedure is that low levels of carbon can be obtained from high carbon charges with comparatively minimal losses of chromium by oxidation. Estimates vary as to the lowest carbon contents that can be achieved, but attainable ones are in the range 0.01–0.02% carbon, as given in Table 1-2, well below the 0.03% maximum for the L grades. Another element whose level can be significantly reduced by AOD is sulfur. Claims have been made to the effect that by the end of 1978, over 90% of the U. S. production of stainless steels will employ AOD. While no figures are currently available, indications are that the AOD melt practice will continue to increase in use.

The third technique listed in Table 1-2, namely, electron beam melting, has been used in the past to produce stainless steels with extremely low interstitial levels. Unlike AOD, this technique requires low carbon charge materials.

Other new trends suggest that there will be increased utilization of chromium plus nickel bearing wastes in the melt charges for stainless steel production. An example is the emergence of new companies, such as the International Metals Reclamation Company, which began constructing facilities in 1978 for converting previously unused stainless steel and other alloy manufacturing and finishing wastes, such as flue dust, grinding swarf, and mill scale, into usable nickel-chromium melt charge.

1-5 THE MORE PREVALENT TYPES OF CORROSION ENCOUNTERED WITH STAINLESS STEELS

The overall cost of corrosion in the U. S. is estimated to be of the order of 4.2% of the gross national product (6). There are no detailed analyses available as to the fraction of this cost that is contributed by corrosion of stainless steel equipment. Some insight is provided, however, about specific industries by published analyses. The review by Collins and Monack describing duPont experience (7) is particularly useful in this regard. This review describes 685 cases of failures of corrosion resistant alloy piping and equipment during the period 1968–1971. Since it is noted that of the duPont purchases of corrosion resistant alloy piping and equipment, more than 90% were in stainless steels (7), this review probably provides the best available insight into the types and frequency of failure of stainless steels that have occurred in the chemical process industry.

When viewed in the context of the severe demands that the chemical

TABLE 1-3. CORROSION FAILURES IN METALLIC PIPING AND EQUIPMENT EXPERIENCE BY duPONT DURING THE PERIOD 1968–1971

Type of Corrosion Failure	1968	1969	1970	1971	Cumulative*
General	18.1	18.1	15.2	9.2	15.2
Stress Corrosion Cracking	13.1	11.5	11.5	15.5	13.1
Pitting	10.1	8.6	8.8	4.2	7.9
Intergranular	7.1	5.3	2.7	9.2	5.6
Erosion-Corrosion	6.1	3.3	3.2	4.2	3.8
Weld Corrosion	0.0	1.9	1.8	6.3	2.5
Corrosion Fatigue	0.0	1.4	2.7	0.7	1.5
High Temperature	0.0	1.9	1.4	1.3	1.3
Crevice	1.0	1.0	0.9	0.7	0.9
Others	5.0	3.5	3.9	3.5	3.4
Total Corrosion	60.5	56.5	52.1	54.8	55.2
Mechanical	39.5	43.5	47.9	45.2	44.8

*685 cases.

process industry places on corrosion resistant materials, it seems noteworthy that only about one-half of the 685 failures could be attributed to corrosion. The specific cumulative percentages quoted, based on the 685 cases, were 55.2% related to corrosion and 44.8% related to mechanical causes. The incidences of failure by corrosion are itemized in Table 1-3 (7).

Even a cursory inspection of the data in Table 1-3 reveals that a serious mode of attack of stainless steels is localized corrosion (i.e., stress corrosion cracking, pitting, intergranular corrosion, corrosion fatigue, and crevice corrosion). These forms of attack accounted for over one-half of the corrosion failures found. Further subdivision emphasizes the importance of stress (stress corrosion cracking, corrosion fatigue) in some of these localized forms of attack (cumulative 14.6% of 685 cases). Such detailed analyses are not available for other industries where the incidence and distribution of the various types of failure may be different than in the chemical industry.

Because of recent conferences held to discuss stress corrosion cracking, this particular type of failure has received an unusual amount of publicity. However, it was estimated that in 1970 industries reporting a

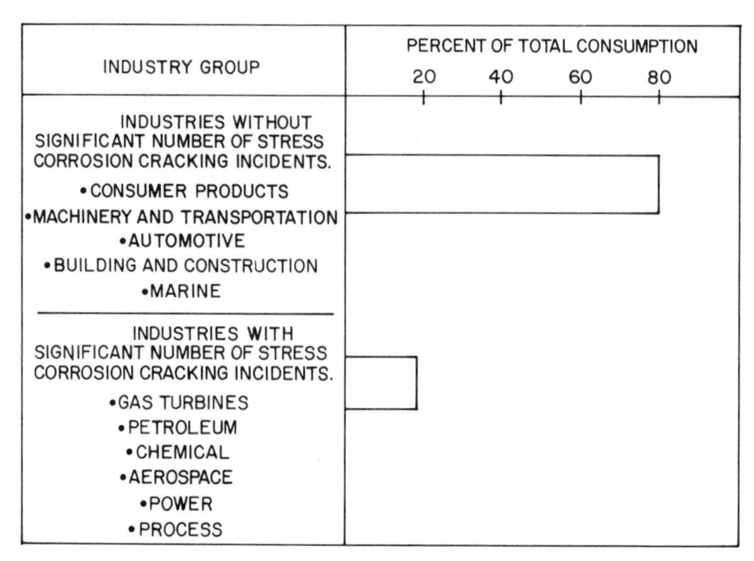

Figure 1-3. Consumption of all grades of stainless steel in 1970 by industries with and without significant stress corrosion cracking incidents.

significant number of stress corrosion failures consumed only some 20% of the stainless steel produced, as indicated in Figure 1-3. It is likely that since that time several of these industries have increased their utilization of more stress corrosion resistant alloys.

In considering the corrosion of stainless steels, it is therefore important also to recognize types of attack other than stress corrosion cracking, such as pitting, crevice corrosion, intergranular corrosion, general corrosion, corrosion fatigue, and attack by high temperature gases. It should be emphasized, however, that numerous failures could be avoided by the correct selection of materials. It is hoped that the information provided in this book will contribute to the achievement of this goal and pinpoint areas where further alloy developments or corrosion research is needed.

REFERENCES

1. W. D. Binder and C. M. Brown, *Proc. Am. Soc. Test. Mater.*, Vol. 46, p. 593, 1946.
2. A. L. Schaeffler, *Met. Prog.*, Vol. 77, No. 2, p. 100B, 1960.
3. H. Schneider, *Foundry Trade J.*, Vol. 108, p. 562, 1960.
4. W. T. DeLong, *Met. Prog.*, Vol. 77, No. 2, p. 100B, 1960.
5. F. B. Pickering, *Int. Met. Rev.*, p. 227, December 1976.
6. *Economic Effects of Metallic Corrosion in the United States*, National Bureau of Standards Special Publication, 1978.
7. J. A. Collins and M. L. Monack, *Mater. Performance*, Vol. 12, p. 11, 1973.

2

COMPOSITION, STRUCTURE, AND MECHANICAL PROPERTIES

2-1 AUSTENITIC STAINLESS STEELS

2-1-1 Nickel Stainless Steels

The AISI 300 Series. As shown in Table 1-1, the 300 series represents by far the largest category of stainless steels produced in the U. S. in the 1970s. The compositions of the steels in the 300 series are shown in Table 2-1, and their mechanical properties are given in Table 2-2. The 300 series represents compositional modifications of the classic 18/8 (18% Cr-8% Ni) stainless steel which has been a popular corrosion resistant material for some 50 years. Among the more important compositional modifications that improve corrosion resistance are (*a*) addition of molybdenum to improve pitting and crevice corrosion resistance, (*b*) lowering carbon content or stabilizing with either titanium or columbium plus tantalum to reduce intergranular corrosion in welded materials, (*c*) addition of nickel and chromium to improve high temperature oxidation resistance and strength, and (*d*) addition of nickel to improve stress corrosion resistance.

Type 304 is the general purpose grade, widely used in applications requiring a good combination of corrosion resistance and formability. Its popularity is indicated by the production figures in Table 1-1. Type 301 exhibits increased work hardening on deformation and is used for higher strength applications. Type 302 is essentially the higher carbon version of type 304 which yields higher strength on cold rolling. Type 302B has a higher silicon content and greater oxidation resistance in high temperature applications.

Types 303 and 303Se contain sulfur and selenium, respectively, and are the free-machining grades used in applications where ease of machining and good surface finish are important. Type 303Se is also used in applica-

TABLE 2-1. COMPOSITIONS OF THE 300 SERIES OF AUSTENITIC STAINLESS STEELS

AISI Grade[a]	Composition[b], %							
	Cr	Ni	C	Mn	Si	P	S	Other
301	16-18	6 - 8	0.15	2.0	1.0	0.045	0.030	--
302	17-19	8 -10	0.15	2.0	1.0	0.045	0.030	--
302B	17-19	8 -10	0.15	2.0	2-3	0.045	0.030	--
303	17-19	8 -10	0.15	2.0	1.0	0.20	0.15[c]	Mo 0.60[d]
303Se	17-19	8 -10	0.15	2.0	1.0	0.20	0.060	Se 0.15[c]
304	18-20	8 -10.5	0.08	2.0	1.0	0.045	0.030	--
304L	18-20	8 -12	0.03	2.0	1.0	0.045	0.030	--
304N	18-20	8 -10.5	0.08	2.0	1.0	0.045	0.030	N 0.10-0.16
305	17-19	10.5-13	0.12	2.0	1.0	0.045	0.030	--
308	19-21	10 -12	0.08	2.0	1.0	0.045	0.030	--
309	22-24	12 -15	0.20	2.0	1.0	0.045	0.030	--
309S	22-24	12 -15	0.08	2.0	1.0	0.045	0.030	--
310	24-26	19 -22	0.25	2.0	1.5	0.045	0.030	--
310S	24-26	19 -22	0.08	2.0	1.5	0.045	0.030	--
314	23-26	19 -22	0.25	2.0	1.5-3.0	0.045	0.030	--
316	16-18	10 -14	0.08	2.0	1.0	0.045	0.030	Mo 2.0 -3.0
316F	16-18	10 -14	0.08	2.0	1.0	0.20	0.10[c]	Mo 1.75-2.50
316L	16-18	10 -14	0.03	2.0	1.0	0.045	0.030	Mo 2.0 -3.0
316N	16-18	10 -14	0.08	2.0	1.0	0.045	0.030	Mo 2-3,N 0.1-0.1
317	18-20	11 -15	0.08	2.0	1.0	0.045	0.030	Mo 3.0 -4.0
317L	18-20	11 -15	0.03	2.0	1.0	0.045	0.030	Mo 3.0 -4.0
321	17-19	9 -12	0.08	2.0	1.0	0.045	0.030	Ti 5xC[c]
330	17-20	34 -37	0.08	2.0	0.75-1.5	0.040	0.030	--
347	17-19	9 -13	0.08	2.0	1.0	0.045	0.030	Cb+Ta 10xC[c]
348	17-19	9 -13	0.08	2.0	1.0	0.045	0.030	Cb+Ta 10xC[c]; Ta 0.1[e]; Co 0.2
384	15-17	17 -19	0.08	2.0	1.0	0.045	0.030	

[a]Source: Source Book on Stainless Steels, American Society for Metals, 1976.
[b]Balance iron. Single values are maximum values unless otherwise noted.
[c]Minimum. [d]Optional. [e]Maximum.
NOTE: ASTM Specification A213-75 designates the carbon content of 304L and 316L as 0.035% (max.).

tions requiring hot upsetting, because of its good hot workability under these conditions.

Type 304L is a lower carbon modification of type 304 used in applications requiring welding. The lower carbon content minimizes carbide precipitation in the heat affected zone near welds, which can lead to intergranular corrosion (weld decay) in certain environments. A further modification is type 304N, to which nitrogen is added to enhance strength.

Types 305 and 384, which have higher nickel contents, exhibit low work hardening rates and are used in applications where cold formability is important. Type 308 is used for welding rods.

Types 309, 310, 314, and 330 have higher nickel and chromium contents to provide oxidation resistance and creep strength at elevated temperatures. Types 309S and 310S are lower carbon versions used to minimize carbide precipitation near welds. Type 330 has a particularly high resistance to carburization and thermal shock.

TABLE 2-2. TYPICAL MECHANICAL PROPERTIES OF THE 300 SERIES OF AUSTENITIC STAINLESS STEELS[a]

AISI Grade	Tensile Strength MPa[c]	Yield Strength (0.2% Offset), MPa	Elongation, %	Hardness, Rockwell B
301	758	276	60	85
302	620	276	50	85
302B	655	276	55	85
303[b]	620	241	50	76
303Se[b]	620	241	50	76
304	586	241	55	80
304L	517	193	55	79
304N	620	331	50	85
305	586	262	50	80
308	586	241	50	80
309	620	310	45	85
309S	620	310	45	85
310	655	310	45	85
310S	655	310	45	85
314	690	345	40	85
316	620	276	50	79
316F	586	241	60	85
316L	517	220	50	79
316N	620	331	48	85
317	620	276	45	85
317L	586	241	55	85
321	620	241	45	80
330[b]	586	310	40	80
347	655	276	45	85
348	655	276	45	85
384	517	241	55	70

[a]Annealed sheet and strip.
[b]Annealed bars.
[c]1 MPa = 145.03 psi.
Source: Source Book on Stainless Steels, American Society for Metals, 1976.

Types 316 and 317 contain molybdenum and have a greater resistance to pitting in marine and chemical industry environments than type 304. For type 316, variations include a low carbon grade (316L), a nitrogen containing grade for increased strength (316N), and a higher sulfur grade for improved machinability (316F).

Types 321, 347, and 348 are grades stabilized with titanium, columbium plus tantalum, and columbium, respectively, for applications requiring welded structures for elevated temperature service. Type 348 is a nuclear grade of stainless steel with a restricted tantalum and cobalt content.

A useful, although somewhat simplified, schematic summarizing the development of the compositional modifications in the 300 series from the basic 18/8 composition is provided by Figure 2-1. Also shown are some of the compositional linkages to other classes of alloys used in certain specialized applications, such as Ni-Cr-Fe alloys, precipitation hardening stainless steels, the austenitic Fe-Ni-Mn-N stainless steels, and the Ni-Cr-Fe-(Mo, Cu, Cb) alloys.

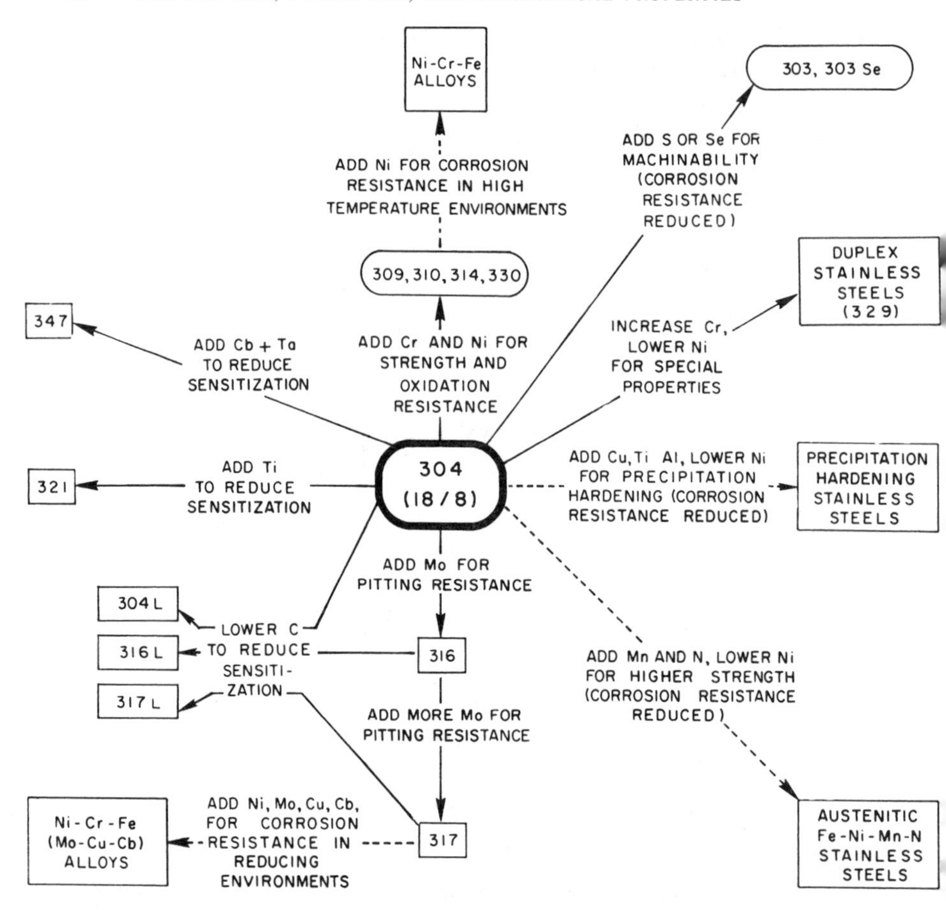

Figure 2-1. Some compositional modifications of 18/8 austenitic stainless steel to produce special properties. Dashed lines show compositional links to other alloy systems.

Delta Ferrite. Relating the compositions of the 300 series of austenitic steels (Table 2-1) to the Schaeffler diagram (Figure 1-2) by calculating the nickel and chromium equivalents shows that the compositions of these steels are balanced to minimize the formation of delta ferrite. This phase is rich in chromium and other ferrite stabilizing elements and lean in nickel and austenite stabilizing elements. It is undesirable from a steel-maker's point of view because it causes difficulty in hot working (1). The presence of delta ferrite is known to decrease pitting resistance. When present as isolated ferrite grains in significant quantities, as in duplex stainless steels, it markedly improves resistance to sensitization and stress corrosion cracking. However, it can decrease resistance to sensitization when present as a continuous grain boundary network. Long term exposure of delta ferrite at elevated temperatures can lead to its transformation to sigma—a hard and brittle phase that can reduce ductility and toughness. (Sigma formation is discussed in Section 2-2 which deals with ferritic stainless steels.)

It is for these reasons that the compositions of the modern 300 series of austenitic stainless steels usually contain sufficient nickel or its equivalents to avoid the presence of significant amounts of delta ferrite. However, some delta ferrite is intentionally allowed to be retained in weld metal and in castings where its presence reduces hot tearing (1), or in special duplex stainless steels (e.g., type 329) in which wear resistance can be improved by intentionally heat treating to transform it to the hard sigma phase (2).

Sensitization. It is evident from Figure 2-1 that in the development of the 300 series, considerable effort has gone into developing grades that resist sensitization. It is now widely accepted that this phenomenon is related to the precipitation of carbide at the austenite grain boundaries. To understand this phenomenon in terms of microstructure, it is instructive to examine the equilibrium relationships and carbon solubility in the 18/8 alloy, illustrated in Figure 2-2 (3). This figure shows that in alloys containing between about 0.03 and 0.7% carbon the equilibrium structure at room temperature should contain austenite, alpha ferrite, and carbide ($M_{23}C_6$). In commercial alloys containing various austenite stabilizers the reaction $\gamma + M_{23}C_6 \rightarrow \gamma + \alpha + M_{23}C_6$ (at line SK) is too sluggish to take place at practical rates of cooling from elevated temperatures. The same applies to the reaction $\gamma \rightarrow \alpha + M_{23}C_6$ at carbon contents below approximately 0.03%. For commercial purity materials the transformation of austenite to alpha ferrite is ignored in practice, and in considering carbon solubility in austenite, a simplified diagram (4), as shown in Figure 2-3, is often considered as being representative of real (i.e., nonequilibrium) situations. In terms of this simplified diagram, austenite containing less than about 0.03% carbon should be stable. Austenite containing carbon in excess of 0.03% should precipitate $M_{23}C_6$ on cooling below the solubility line. However, at relatively rapid rates of cooling this reaction is partially suppressed. This is the case in practice when 18/8 steel containing more than 0.03% carbon is heat treated at 1050°C, to remove effects of cold work or hot working, and cooled at a fairly rapid rate to room temperature. While some carbide may have precipitated on cooling, the room temperature austenite is still largely supersaturated with respect to carbon.

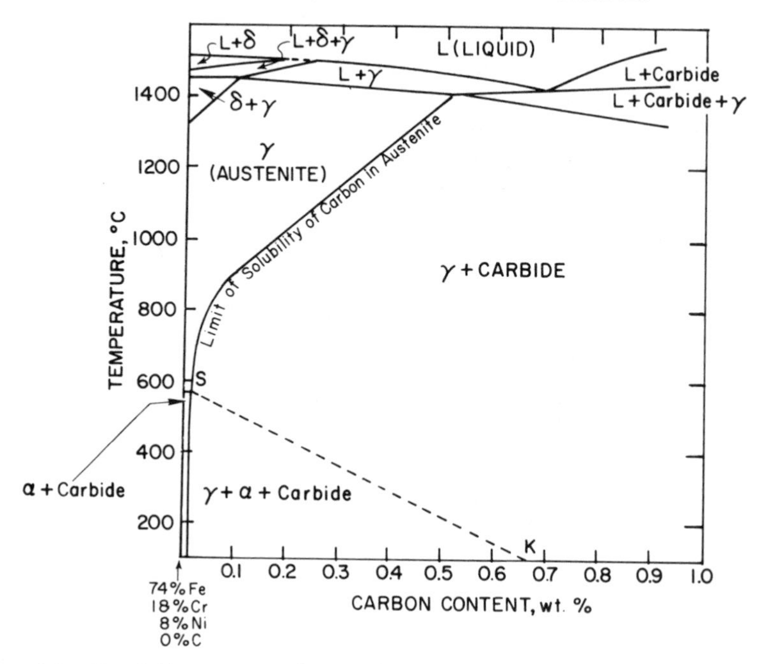

Figure 2-2. Pseudobinary phase diagram for a Fe-18% Cr-8% Ni alloy with varying carbon content. (After Krivobok.)

If this supersaturated austenite is reheated to elevated temperatures within the γ + $M_{23}C_6$ field, further precipitation of the chromium rich $M_{23}C_6$ will take place at the austenite grain boundaries. Certain time-temperature combinations will be sufficient to precipitate this chromium rich carbide, but insufficient to rediffuse chromium back into the austenite near the carbide. This will result in the formation of envelopes of chromium depleted austenite around the carbide (5, 6), which in certain environments will not be resistant to corrosive attack. Since the carbides precipitate along grain boundaries, the net effect is grain boundary corrosion. This attack, when caused by welding, has been known as "weld decay," and has been extensively studied [e.g., (7)]. Modern practice is to use the term "sensitized" to describe chromium depletion irrespective of whether it has resulted from slow cooling, heat treatment, elevated temperature service, or welding.

The chromium depletion theory is now widely accepted to explain sensitization. However, other theories have been proposed in the past. These have been recently reviewed and discussed in detail by Henthorne (8).

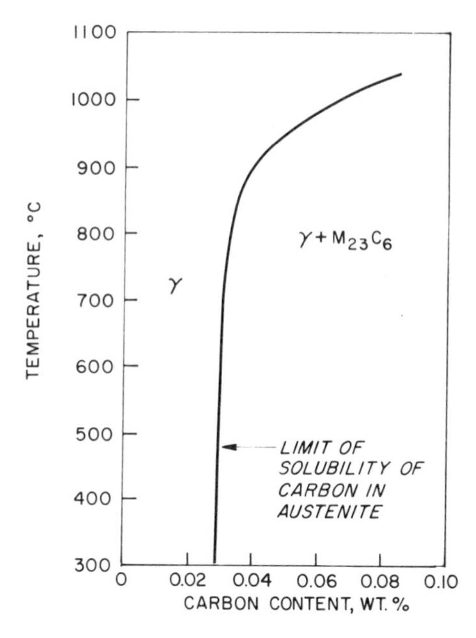

Figure 2-3. Solid solubility of carbon in a Fe-18% Cr-8% Ni alloy. (After Husen and Samans.)

The current metallurgical remedies used to reduce sensitization are consistent with the chromium depletion theory and include (*a*) the use of low carbon (0.03% maximum) grades of stainless steel (i.e., types 304L, 316L, and 317L), (*b*) postweld heat treatment to rediffuse chromium back into the impoverished austenite, and (*c*) the use of titanium additions (type 321) or columbium plus tantalum additions (type 347) to precipitate the carbide at higher temperatures so that little carbon is left to precipitate as the chromium rich grain boundary carbide during cooling. Types 321 and 347 are sometimes given a stabilizing treatment at 900–925°C to ensure maximum precipitation of carbon as titanium or columbium carbides. All these remedies have certain advantages and disadvantages. Thus the low carbon grades have slightly lower strength, postweld heat treatment is not always practicable in large structures, and the stabilized grades can suffer another form of corrosive attack known as knife-line attack. The latter is discussed in Chapter 6.

Martensites Induced by Strain or Subzero Cooling. Most of the 300 series stainless steels can undergo martensitic transformations, $\gamma \rightarrow \epsilon$ and $\gamma \rightarrow \alpha'$, as a result of cold working, particularly below room temperature [e.g., (9)]. Steels containing 8–10% nickel harden on plastic deformation not only as a result of strain hardening, but also as a result of these strain induced transformations. Figure 2-4 summarizes the effect of cold rolling

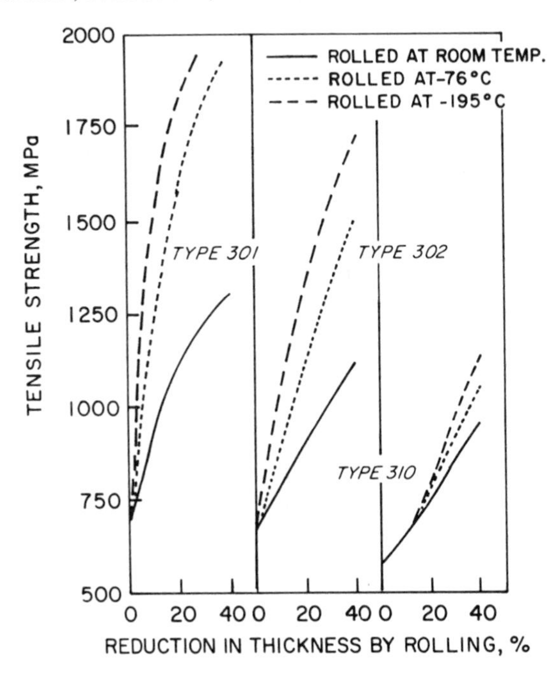

Figure 2-4. Effect of rolling temperature on tensile strength of type 301, 302, and 310 stainless steels. (After Krivobok and Talbot.)

temperature, amount of plastic deformation, and alloy type on hardening (10). The more highly alloyed type 310 stainless steel shows little difference in hardening rate as a function of rolling temperature, suggesting that in this steel the austenite is on the verge of being stable. The epsilon phase has a hexagonal close packed lattice, whereas the alpha prime martensite is body centered cubic (9). Their formation is favored by the electrolytic charging of hydrogen into the austenite (11). The formation of these phases has been linked with some theories attempting to explain the mechanism of stress corrosion cracking of austenitic stainless steels [see (12) for a review].

Sulfide Inclusions. Sulfur is present in austenitic stainless steels as an impurity (0.03%) or as an intentional addition to improve machinability (e.g., type 303) usually at a level of 0.3%. Since the solubility of sulfur in stainless steels is less than 0.01% at room temperature, it usually exists as metal sulfide. The sulfides are predominantly manganese sulfides, but they also contain chromium and other elements. They are generally detrimental to pitting resistance in all grades of stainless steels, not just austenitics, although the extent to which they act as pit sites depends on their manganese and chromium content and chemical treatments aimed at removing them from the surface layers (13).

It should be noted that the production of low sulfur material does not present a problem in AOD refining. It is estimated that sulfur contents as low as 0.006–0.01% can be obtained with AOD practice. However, since it has not been demonstrated that variations in sulfur contents in the range 0.006–0.03% significantly affect pitting resistance, there has been no incentive to produce low sulfur material. As discussed in Chapter 4, manganese content appears far more important. In fact too low a sulfur content can adversely affect machining and cutting properties, and sulfur is often back-added to the usual levels (0.03% maximum) to ensure adequate machinability of regular grades, such as type 304.

2-1-2 Nickel-Manganese-Nitrogen Stainless Steels

These austenitic steels are designated by AISI as the 200 series. Their compositions are shown in Table 2-3 and their mechanical properties are given in Table 2-4. As indicated by Table 1-1, only types 201 and 202 are currently produced in significant quantities. In the United States these steels became popular during the early 1950s which was a period of scarcity of nickel, and during which additions of austenite stabilizers other than nickel were evaluated. These other austenite stabilizers are carbon, nitrogen, manganese, copper, and cobalt. As single additions, none of these elements is satisfactory in producing an austenitic stainless steel. As pointed out by U. S. Steel Corporation (14), carbon in amounts

TABLE 2-3. COMPOSITIONS OF THE 200 SERIES OF AUSTENITIC STAINLESS STEELS

AISI	Composition[a], %								
Grade	Cr	Ni	C	Mn	Si	P	S	N	Other
201	16 -18	3.5-5.5	0.15	5.5- 7.5	1.0	0.060	0.030	0.25	-
202	17 -19	4-6	0.15	7.5-10	1.0	0.060	0.030	0.25	-
205	16.5-18	1-1.75	0.25	14-15.5	1.0	0.060	0.030	0.40	1-1.75Mo
216[b]	17.5-22	5-7	0.08	7.5-9	1.0	0.045	0.030	0.50	3Mo
216L[b]	17.5-22	5-7	0.03	7.5-9	1.0	0.045	0.030	0.50	3Mo

[a] Balance iron. Single values are maximum values unless otherwise noted.
[b] Not standard AISI grades.
Sources: Source Book on Stainless Steels, American Society for Metals, 1976.
Allegheny Ludlum Alloy Data.

TABLE 2-4. TYPICAL MECHANICAL PROPERTIES OF THE 200 SERIES OF AUSTENITIC STAINLESS STEELS[a]

AISI Grade	Tensile Strength MPa[b]	Yield Strength (0.2% Offset) MPa	Elongation %	Hardness, Rockwell B
201	758	379	55	90
202	689	379	55	90
205	862	482	58	98
216[c]	689	379	45	92
216L[c]	689	379	45	92

[a]Annealed sheet and strip.
[b]1 MPa = 145.03 psi.
[c]Not standard AISI grades.
Sources: Source Book on Stainless Steels, American Society for Metals, 1976.
Allegheny Ludlum Alloy Data.

necessary to form a completely austenitic structure has a detrimental effect on ductility and corrosion resistance, nitrogen cannot be added in sufficiently large quantities, manganese even in amounts above 25% will not form completely austenitic structures in alloys containing over 15% chromium, copper has a detrimental effect on hot ductility, and cobalt is an expensive alloying addition.

Although a nickel-free proprietary austenitic stainless steel, Tenelon (Fe-18% Cr-15% Mn-0.4% N-0.1% C), was developed during the previously noted period, a subsequently modified alloy, Cryogenic Tenelon, contains 5.5% nickel (15). Similarly, in types 201 and 202 only about 4% nickel is replaced by 7% manganese and 0.25% nitrogen. Types 201 and 202 have higher yield strengths than their corresponding 300 series counterparts (Table 2-4 vs. Table 2-2), but their overall corrosion resistance is generally considered inferior to that of the 300 series of steels.

A higher molybdenum grade, type 216, is also available that contains up to 0.5% nitrogen, and that is claimed to have good pitting and crevice corrosion resistance (16).

The production of high manganese alloys can produce problems associated with attack on furnace linings by the generation of excessive amounts of manganese oxide in the slag. The production of the 200 grades has, therefore, not been a popular activity with steelmakers. Regarding nitrogen, it is known that in AOD refining this element can be removed to levels of about one-third of that obtained in conventional electric arc melting. However, nitrogen can also be easily back-added as gaseous nitrogen to AOD refined heats to relatively high and reproducible levels.

2-1-3 Duplex Stainless Steels

As can be seen from the Schaeffler diagram (Figure 1-2), steels containing about 28% chromium and 6% nickel contain both austenite and ferrite and hence are known as duplex. The exact amounts of each phase can be varied by the introduction of other austenite and ferrite stabilizers. The compositions of some of the currently available duplex stainless steels are shown in Table 2-5 and their mechanical properties are listed in Table 2-6. Only type 329 is an AISI grade; the others are proprietary grades.

As discussed in Chapter 7, the duplex structure renders the propagation of stress corrosion cracks more difficult. Therefore, in the annealed condition duplex stainless steels are regarded as more stress corrosion resistant than some of the lower alloy austenitic grades. The duplex structure is also more resistant to sensitization, but less resistant to crevice corrosion and pitting.

In general, hot working difficulties have retarded the development of

TABLE 2-5. COMPOSITIONS OF SOME DUPLEX STAINLESS STEELS

Designation	Typical Composition[a], (%)							
	Cr	Ni	C	Mn	Si	P	S	Other
AISI Type 329	28.0	6.0	0.10	2.00	1.0	0.04	0.03	Mo 1.5
326[b]	26.0	6.5	0.05	1.00	0.6	0.01	0.01	Ti 0.25
SANDVIK 3RE60	18.5	4.5	0.02	1.50	1.6	0.01	0.01	Mo 2.5

[a]Balance iron.
[b]Earlier version known as developmental alloy IN-744; commercially available since 1970 under proprietary designations of UNILOY 326, AL 326 and H-326.
Sources: Various.

TABLE 2-6. TYPICAL MECHANICAL PROPERTIES OF SOME DUPLEX STAINLESS STEELS[a]

Designation	Tensile Strength MPa[b]	Yield Strength (0.2% Offset) MPa	Elongation, %	Hardness, Rockwell B
AISI Type 329	724	551	25	98
326	689	517	35	95
SANDVIK 3RE60	717	482	48	92

[a]Mill annealed condition.
[b]1 MPa = 145.03 ksi.
Sources: Various.

the duplex stainless steels. It is a point of interest, therefore, that relatively recently a duplex steel designated by several U. S. suppliers as "326" has been made available (15), which is reported to be readily workable in the 870–980°C range due to microstructural control (17).

2-1-4 Higher Alloys

It is important from the viewpoint of corrosion resistance to describe also some of the more highly alloyed austenitic materials that contain less than 50% iron. It is convenient to divide these higher alloys into two groups, namely, Ni-Cr-Fe alloys and Ni-Cr-Fe-(Mo, Cu, Cb) alloys.

The Ni-Cr-Fe alloys can be regarded as higher alloy extensions of type 304, and their nominal compositions are shown in Table 2-7. While Inconel alloy 600 is a general purpose corrosion resistant material for elevated temperatures, the development of light water cooled nuclear power plants has placed it in the position of being the preferred material for steam generating applications. Alloy 800 has been used in Europe, and Inconel alloy 690 is a new material currently being evaluated for steam generator applications. The compositions shown in Table 2-7 are typical of those used in pressurized water reactor steam generators. Higher carbon versions are used in other applications.

The Ni-Cr-Fe-(Mo, Cu, Cb) alloys, given in Table 2-8, are used in applications requiring corrosion resistance to reducing acids, pitting, and crevice corrosion. These materials can be regarded as higher alloy extensions of type 317. Inconel alloy 625 and Hastelloy alloy C-276 are among

TABLE 2-7. COMPOSITIONS OF SOME AUSTENITIC Ni-Cr-Fe ALLOYS

Alloy	Typical Composition[a], (%)					
	Ni	Cr	Fe	C	Co	Ti
AISI Type 304[b]	10	19	69	0.06	0.15	0.008
Alloy 800[c]	33	21	38	0.03 max.	0.10 max.	0.60 max.[f]
INCONEL Alloy 600[d]	76	15	8	0.04	0.03	0.25
INCONEL Alloy 690[e]	60	30	9.5	0.03	0.03	0.25

[a]Typical compositions for tubes and pipes for high temperature water service in light water cooled nuclear power plants.
[b]Stainless steel used for pipes in boiling water reactors.
[c]Alloy used in West Germany for steam generator tubing.
[d]Alloy used in the U.S. for steam generator tubing.
[e]New alloy being evaluated for possible use for steam generator tubing.
[f]Ti/C = 12 min.: Ti/(C+N) = 8 min.
Sources: Various.

TABLE 2-8. COMPOSITIONS OF SOME AUSTENITIC Ni-Cr-Fe-(Mo, Cu, Cb) ALLOYS

Alloys	Typical Composition (Wt. %)							
	Ni	Cr	Fe	Mo	C	Cb	Cu	Other
AISI Type 317	13.0	19.0	64.0	3.5	0.05	--	--	--
SANDVIK 2RN65	24.0	17.5	46.0	4.7	0.02	--	--	--
SANDVIK 2RK65	25.0	19.5	50.0	4.5	0.02	--	1.5	--
AL-6X	25.0	20.0	48.5	6.5	0.01	--	--	--
JESSOP 700	25.0	20.0	49.0	4.4	0.02	--	--	--
JESSOP 777	25.0	21.0	45.0	4.6	0.02	0.3	2.5	--
UDDEHOLM 904L	25.0	20.0	49.0	4.5	0.02	--	1.5	--
HAYNES No. 20 (Mod)	26.0	22.0	45.0	5.0	0.03	--	--	Ti 4xC
CARPENTER 20Cb-3	33.0	20.0	41.0	2.0	0.03	0.8	3.0	--
INCOLOY Alloy 825	42.0	20.0	30.0	3.0	0.03	--	2.0	--
HASTELLOY Alloy G	45.5	22.0	19.5	6.5	0.03	2.0	2.0	W 0.5
HASTELLOY Alloy C-276	59.0	16.0	5.0	16.0	0.02	--	--	W 4.0
INCONEL Alloy 625	62.0	22.0	3.0	9.0	0.05	3.5	--	--

Source: Supplier literature.

the most resistant austenitic alloys currently available for severe aqueous environments.

In terms of the Schaeffler diagram, the compositions of these higher alloys lie deep within the region of austenite stability, which ensures freedom from delta ferrite and strain induced martensite. However, as in the case of the austenitic stainless steels, at elevated temperatures these alloys can precipitate grain boundary carbides, and in the case of Ni-Cr-Fe-(Mo, Cu, Cb) alloys, more complex intermetallic compounds may be precipitated. Accordingly, the compositions of these proprietary alloys are adjusted to minimize such precipitation during welding or high temperature service. For specific information regarding these alloys the reader should consult the supplier literature.

2-2 FERRITIC STAINLESS STEELS

2-2-1 The AISI 400 Series

The compositions of the 400 series of AISI standard grades considered to be ferritic are shown in Table 2-9 and their mechanical properties are

TABLE 2-9. COMPOSITIONS OF THE 400 SERIES OF FERRITIC STAINLESS STEELS

AISI Grade	Composition[a], (%)						
	Cr	C	Mn	Si	P	S	Other
405	11.5 -14.5	0.08	1.0	1.0	0.040	0.030	Al 0.10-0.30
409	10.5 -11.7	0.08	1.0	1.0	0.045	0.045	Ti 6xC[d],0.75[c]
429	14 -16	0.12	1.0	1.0	0.040	0.030	--
430	16 -18	0.12	1.0	1.0	0.040	0.030	--
430F	16 -18	0.12	1.25	1.0	0.060	0.15[d]	Mo 0.6[e]
430FSe	16 -18	0.12	1.25	1.0	0.060	0.060	Se 0.15[d]
434	16 -18	0.12	1.0	1.0	0.040	0.030	Mo 0.75-1.25
436	16 -18	0.12	1.0	1.0	0.040	0.030	Cb+Ta 5xC[d],0.7[c],Mo 0.75-1.25
439[b]	17.75-18.75	0.07	0.60	0.60	0.040	0.030	Ti 12xC[d],1.0[c],Ni 0.5
442	18 -23	0.20	1.0	1.0	0.040	0.030	--
446	23 -27	0.20	1.5	1.0	0.040	0.030	N 0.25

[a]Balance iron. Single values are maximum values unless otherwise noted.
[b]439 is not a standard AISI grade.
[c]Maximum.
[d]Minimum.
[e]Optional.
Source: Source Book on Stainless Steels, American Society for Metals, 1976.

listed in Table 2-10. The Schaeffler diagram (Figure 1-2) indicates that a predominantly ferritic structure should be obtained if the chromium content exceeds approximately 12%, particularly in the presence of other ferrite stabilizers. Thus, in type 405, the addition of approximately 0.2% aluminum (a ferrite stabilizer) and maintenance of carbon (an austenite stabilizer) to relatively low levels (about 0.05%) ensures a largely ferritic structure. As can be seen from Table 1-1, by far the most popular ferritic grade has been type 430, which is the basic 17% Cr stainless steel. Its free-machining modifications, types 430F and 430FSe, have also been used to some extent (Table 1-1). In the past, type 430 was the multipurpose ferritic stainless steel, with a range of chromium content between 14 and 18% giving the user a choice of properties. Specifying chromium on the low side improved weldability, impact resistance, strength, and hardness, but with some sacrifice in corrosion resistance. With chromium on the high side, there was a gain in corrosion resistance, particularly in nitric acid, but loss in mechanical properties, particularly impact strength. More recent specifications have narrowed the chromium range to between 16 and 18%. Types 409 and 439 are stabilized with titanium to improve

TABLE 2-10. TYPICAL MECHANICAL PROPERTIES OF THE 400 SERIES OF FERRITIC STAINLESS STEELS[a]

AISI Grade	Tensile Strength MPa[b]	Yield Strength MPa	Elongation, %	Hardness, Rockwell B
405	482	276	30	80
409	469	276	25	75
429	486	293	30	76
430	517	310	30	82
430F	551	379	25	86
430FSe	551	379	25	86
434	531	365	23	83
436	531	365	23	83
439[c]	482	280	30	80
442	551	345	20	90
446	551	345	25	86

[a]Annealed material.
[b]1 MPa = 145.03 psi.
[c]439 is not a standard AISI grade.
Source: Source Book on Stainless Steels, American Society for Metals, 1976.

corrosion resistance at welds in mild environments. Recently, type 409 has been produced in large quantities for automobile emission control equipment, as indicated in Table 1-1.

Type 446 is the highest chromium grade in the ferritic AISI 400 series and has the highest corrosion and oxidation resistance of this series. Nitrogen, columbium, aluminum, and titanium can be added to restrict grain growth.

The ferritic structure in these steels introduces a number of complications of a metallurgical nature which can influence corrosion behavior. Among the metallurgical problems encountered in ferritic stainless steels are the ductile-to-brittle transition, 475°C embrittlement, sigma phase formation, high temperature embrittlement, low ductility of the welded condition, and sensitization.

The Ductile-to-Brittle Transition. Toughness of stainless steels is usually measured with either Izod or Charpy impact test specimens at various test temperatures. Such tests yield a parameter known as the ductile-to-brittle transition temperature (DBTT), below which a given material has

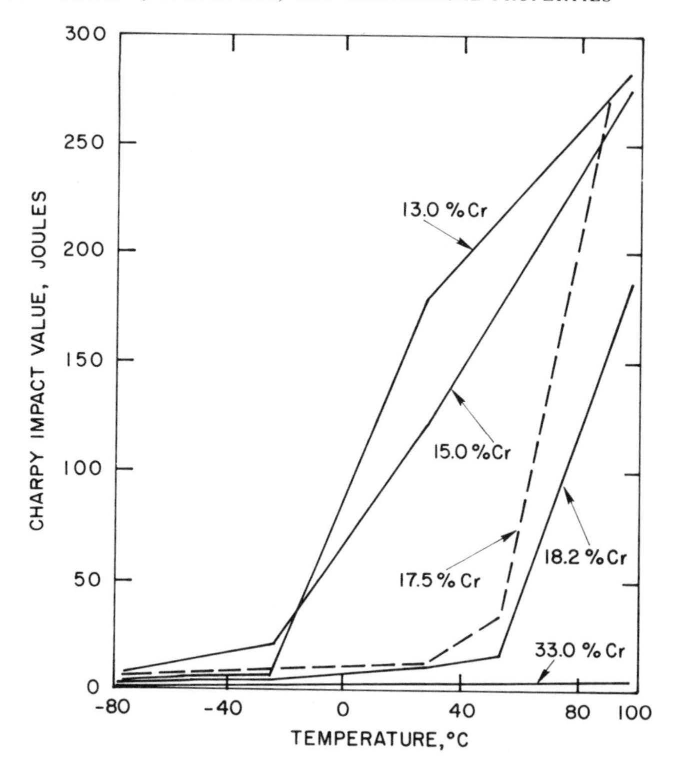

Figure 2-5. Effect of test temperature on the notch toughness of iron-chromium alloys containing about 0.01% carbon. The 33% chromium alloy contained 0.07% carbon. (After Krivobok.)

low impact strength. In the case of the ferritic stainless steels, much of the early investigation of the parameters affecting DBTTs were carried out by Krivobok (18), such as establishing the fact that the DBTT is increased with increasing chromium content of the ferritic alloy. This is illustrated in Figure 2-5 in the case of low carbon (0.01%) iron-chromium alloys containing various amounts of chromium. Various DBTTs for commercially produced 400 series of alloys are given in ASM's *Metals Handbook* (19).

Several factors enter into determining the DBTT, including the thickness of the material, grain size, and the presence and quantity of certain alloying elements. The thickness of the material appears to be a strong factor, as shown in Figure 2-6 for types 409, 439, and some proprietary developmental ferritic stainless steels. In this figure the curves shown for types 409 and 439 are the upper limit of a broad scatter-band (20). The dependence of the DBTT on material thickness is thought to be associated

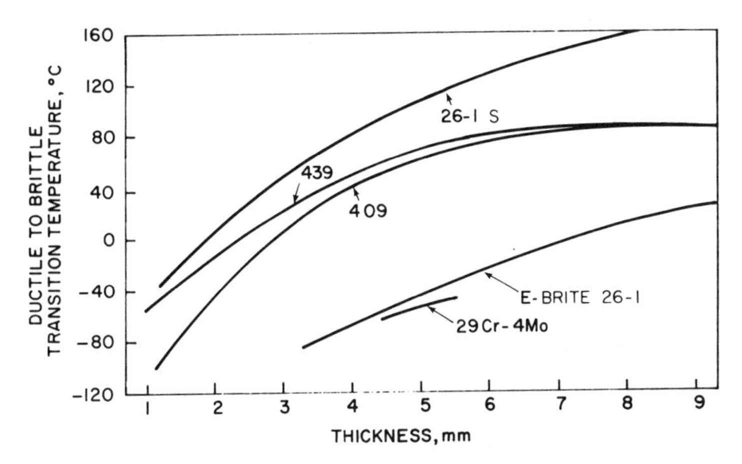

Figure 2-6. Effect of material thickness on the ductile-to-brittle transition temperature of ferritic stainless steels. (After Lula.)

with constraints in thicker materials that prevent deformation through the thickness.

Grain size is also important in determining the DBTT, with smaller grain sizes favoring lower DBTTs (1). Since the grain size of the fully ferritic stainless steels can only be reduced by cold rolling and recrystallizing at selected temperatures, thinner gage material, which receives more cold reduction in the forming operation, generally has smaller grains and a lower DBTT. Among alloying elements that raise the DBTT are titanium, carbon, and nitrogen. Another factor which is reported to lower DBTT is increased cooling rate (20). This effect increases with increasing chromium content and appears to be related to reduced residence time, during cooling, at temperatures causing 475°C embrittlement.

475°C Embrittlement. When ferritic stainless steels containing more than about 12% chromium are heated to temperatures above 340°C, notch ductility is considerably reduced and the material becomes brittle. The maximum effect occurs at approximately 475°C, hence the phenomenon is known as 475°C embrittlement. Generally, the higher the chromium content of the alloy, the shorter the time required to develop this embrittlement. The cause of 475°C embrittlement has been recognized as the precipitation of a chromium rich alpha prime (α')* phase (21). The pre-

*This phase should not be confused with alpha prime martensite that can form in certain austenitic stainless steels on cold working.

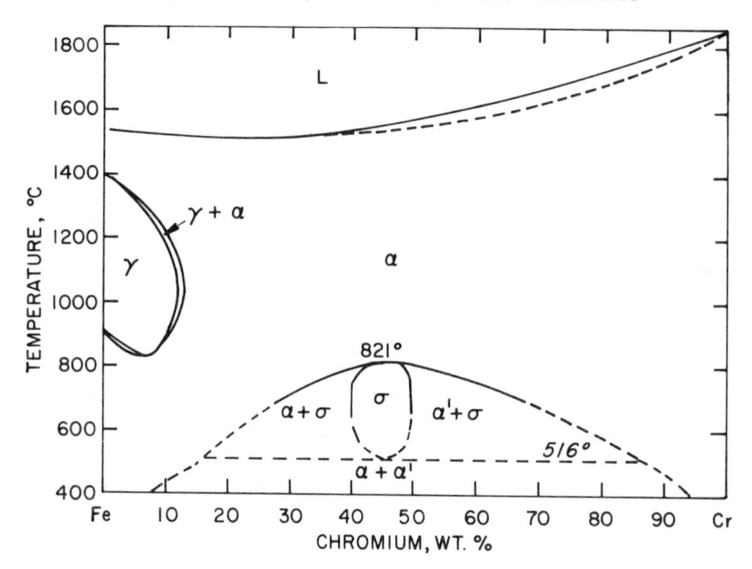

Figure 2-7. Iron-chromium equilibrium diagram incorporating modifications by Williams.

cipitation of this phase is due to the decomposition of the alpha (α) iron-chromium solid solution which is unstable at lower temperatures.

To understand the composition-structure-temperature relationships governing the precipitation of α', it is instructive to refer to the iron-chromium equilibrium diagram, given in Figure 2-7, which incorporates the features proposed by Williams (22). According to this diagram 475°C embrittlement may be expected at temperatures below 516°C over a wide range of chromium contents. The lower temperature limit, usually considered to be about 340°C, is determined by the kinetics of the nucleation and growth of α' (23). Long term exposures (e.g., 100 hours) at temperatures above 516°C (e.g., at 538°C) leads to another form of embrittlement, which is not caused by precipitation of α', but by the precipitation of carbonitrides (23).

The phenomenon of 475°C embrittlement can be removed by reheating to temperatures above 516°C (e.g., 600°C) and rapidly cooling to room temperature. Prolonged heating at temperatures in the $\alpha + \sigma$ range, shown in Figure 2-7, should be avoided, however, because of the possibility of sigma (σ) formation.

Sigma Phase Embrittlement and High Temperature Embrittlement. Figure 2-7 shows the region of temperatures and compositions over which sigma is an equilibrium phase. Sigma forms very slowly, first developing at grain boundaries, and is a hard and brittle intermetallic readily identifiable by microscopy and x-ray diffraction. Its formation is favored by high chromium contents, molybdenum, and cold work (4). The presence of the sigma phase increases hardness, but it decreases ductility, notch toughness, and corrosion resistance. However, because of its slow rate of formation, sigma is usually a service problem where long exposures at elevated temperatures are involved. Sigma can be redissolved by heating to temperatues of 900°C or above, followed by rapid cooling to avoid 475°C embrittlement.

However, heating high chromium ferritic stainless steels containing moderate to high levels of interstitial elements to 1000°C and above can result in extreme loss in toughness and ductility at room temperature. This effect, termed high temperature embrittlement, has been reviewed by Demo (24) and is thought to be caused by the precipitation of carbonitrides.

It should be noted that sigma formation is possible between 565 and 925°C also in austenitic stainless steels containing more than 16% chromium and less than 32% nickel (25).

Weld Ductility. A major difficulty associated with the welding of ferritic stainless steels is lack of ductility of the welded region. Four factors can contribute to this lack of ductility, namely, grain coarsening, 475°C embrittlement, high temperature embrittlement by carbonitrides, and to a lesser extent the formation of sigma phase. These four factors are favored by increasing chromium content of the ferritic steels. Hence weld ductility and corrosion resistance are inversely related, as shown in Figure 2-8 (26). Thus the lower chromium grades could be more easily welded, but their lower chromium content would reduce corrosion resistance. Type 430 can be welded with a matching electrode, but usually preweld and postweld heat treatments are recommended (27). The highest chromium grades, types 442 and 446, being fully ferritic are particularly susceptible to high temperature embrittlement (24) and to ferrite grain growth in the weld and heat-affected zones (1) and hence exhibit low ductility. The ferrite grain size cannot be refined by high temperature phase transformation since their compositions lie outside the γ loop, as illustrated in Figure 2-7.

In some instances austenitic electrodes (types 308, 309, and 310) are used for the welding of types 430 and 446. With the austenitic fillers no preheating is necessary if the ferritic steel is above 16°C (27). However, while the austenitic fillers may produce welds of greater ductility and toughness, the presence of a dissimilar metal may adversely affect corrosion and stress corrosion resistance or lead to other problems arising from different coefficients of thermal expansion.

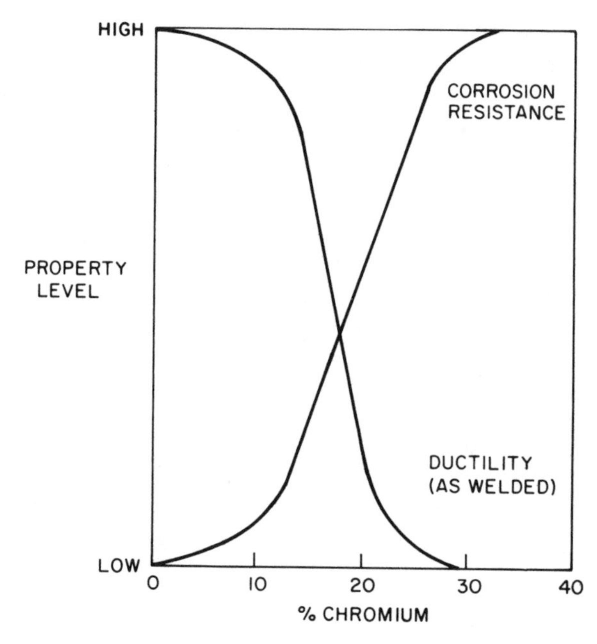

Figure 2-8. Effect of chromium on properties of ferritic stainless steels.

Sensitization. Ferritic stainless steels, like the austenitic grades, can exhibit susceptibility to intergranular corrosion after certain thermal treatments. However, this subject has received much less study than the comparable phenomenon in austenitics because of the much smaller usage of ferritics. In the case of the ferritics, most of the studies have been concerned with type 430.

In attempting to understand the metallurgical structure which is susceptible to intergranular attack, it is instructive to examine Figure 2-9 (3), which illustrates the phase relationships and carbon solubility in a Fe-18% Cr alloy. This figure shows that the alpha ferrite has a low solubility of carbon, which if present in any significant quantities in solid solution at elevated temperatures should precipitate as carbide on cooling. It is generally accepted that to sensitize type 430, the alloy must be heated to temperatures at which austenite forms,* that is, above line *P-L* in Figure 2-9. Material which has been heated into this temperature range shows severe susceptibility to intergranular attack (even in water), irrespective

*Higher chromium ferritics, which do not have a ferrite to austenite transition, will also sensitize due to changes in carbon and nitrogen solubility with temperature (see Chapter 6).

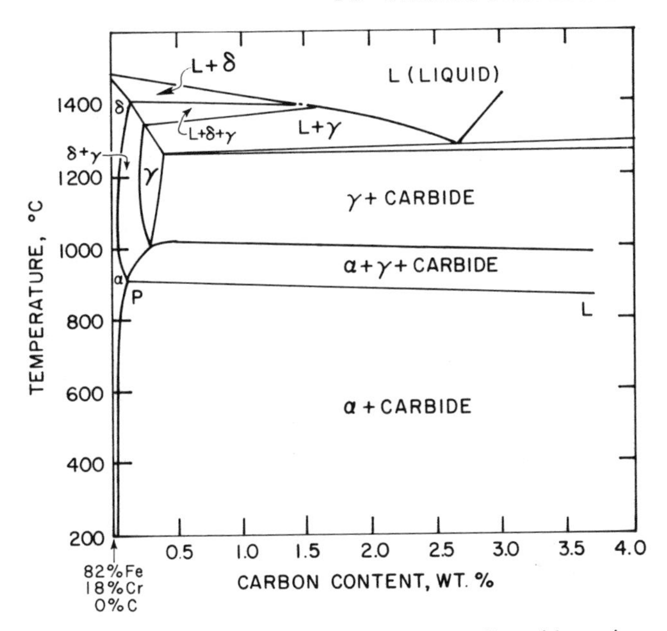

Figure 2-9. Pseudobinary phase diagram for a Fe-18% Cr alloy with varying carbon content. (After Krivobok.)

of whether it is water quenched or air cooled to room temperature. Thus welds and portions of the heat-affected zone that have experienced temperatures above line *P-L* in Figure 2-9 are susceptible to intergranular attack. Remedial measures used for austenitics, such as lowering carbon to 0.03% maximum, do not eliminate the problem, and neither do the additions of stabilizing elements, although the latter minimize the effect (8). The best remedy appears to be to reanneal the sensitized material at approximately 800°C.

It is clear from the foregoing, therefore, that the temperature regimes for sensitization and its removal are considerably different for austenitic and ferritic stainless steels. Nevertheless, the more recent studies of the sensitization of ferritic stainless steels support the idea that the mechanism of intergranular attack in both ferritic and austenitic stainless steels derives from the same basic phenomenon—chromium depletion of the matrix in the vicinity of precipitated carbides and nitrides at the grain boundaries (28–32). Essentially, the proponents of the chromium depletion theory argue that heating above line *P-L* in Figure 2-9 dissolves some of the carbides and nitrides into the ferrite and austenite. The ferrite becomes saturated with carbon and nitrogen. On cooling the solubility of carbon and nitrogen is greatly lowered, and carbides and nitrides will precipitate very rapidly. Carbides and nitrides formed at lower temperatures (400–700°C) will cause chromium depletion in their vicinity. Holding at higher temperatures (e.g., about 800°C) will cause chromium to rediffuse back into the depleted regions. Thus the difference in the sensitization temperature regimes between austenitics and ferritics is explained in terms of the much lower solubilities of carbon and nitrogen in ferrite than in austenite, and the consequent more rapid precipitation of carbides and nitrides from ferrite.

It should be noted that the chromium depletion theory has not gained the level of acceptance for ferritic stainless steels that it has for austenitics. Theories other than those based on chromium depletion have been proposed for ferritic stainless steels. For a comprehensive summary and critique of these other theories the reader is referred to the study by Henthorne (8). However, because the chromium depletion theory explains both sensitization and measures to avoid it in terms of a simple and readily understandable metallurgical mechanism which is applicable to all classes of stainless steel, it is considered to be the most feasible theory.

2-2-2 New High Chromium, Low Interstitial Ferritic Stainless Steels

In the last decade in the U. S. there has been a significant research and development effort to introduce high chromium, low interstitial ferritic stainless steels containing molybdenum. This activity has been closely linked with the development and introduction of two other relatively new steel production techniques known as electron beam melting and large volume vacuum induction melting. Some of the developmental ferritic stainless steels produced by these techniques are described in terms of composition and mechanical properties in Tables 2-11 and 2-12, respectively. The essential features of these materials are the high chromium contents and the very low carbon and nitrogen levels, particularly in the electron beam melted material. It has been reported that to avoid sensitization, % C + % N should not exceed 0.01% for 26-1 and 0.025% for 29-4.*

The very low carbon and nitrogen level of E-Brite 26-1 keeps the ductile-to-brittle transition temperature to below room temperature over a wide range of thickness, as shown in Figure 2-6 (20). However, because of the low interstitial levels extreme care must be taken in welding to avoid the pickup of contaminants from the atmosphere. The higher carbon titanium stabilized grade 26-1S is supplied only in light gages because of

**Materials Engineering*, Manual 267, *Ferritic Stainless Steels*, p. 69, April 1977.

TABLE 2-11. COMPOSITIONS OF SOME HIGH CHROMIUM, LOW INTERSTITIAL FERRITIC STAINLESS STEELS MADE BY NEW PRODUCTION TECHNIQUES

Designation	Composition[a], (%)							
	Cr	C	Mn	Si	P	S	Mo	Other
E-BRITE 26-1[b]	26	0.001	0.01	0.25	0.01	0.01	1.0	N 0.010
26-1S	25-27	0.06	0.75	0.75	--	--	0.75-1.50	Ti 0.2-1.0,7 x (C+N) min, N 0.04 max.
29-4[c]	29	0.005	0.10	0.02	0.015	0.01	4.0	N 0.013
29-4-2	29	0.005	0.10	0.02	0.015	0.01	4.0	Ni 2.0,N 0.013

[a]Balance iron. Single values are maximum values.
[b]Produced by electron beam melting.
[c]Produced by large volume vacuum induction melting.

Source: Materials Engineering, April, 1977.

TABLE 2-12. TYPICAL MECHANICAL PROPERTIES OF HIGH CHROMIUM, LOW INTERSTITIAL STAINLESS STEELS MADE BY NEW PRODUCTION TECHNIQUES

Designation	Tensile Strength MPa[b]	Yield Strength MPa	Elongation,%	Hardness, Rockwell B
E-BRITE 26-1	500	345	35	85
26-1S	517	358	30	85
29-4	620	517	25	95
29-4-2	655	586	22	97

[a]Annealed materials.
[b]1 MPa = 145.03 psi.

Source: Materials Engineering, April, 1977.

its higher ductile-to-brittle transition temperatures in thicker sections (Figure 2-6).

The developmental alloys designated as 29-4 and 29-4-2 also rely on their low interstitial levels to prevent sensitization during welding, but to avoid contamination they must be welded with the same extreme care as E-Brite 26-1, requiring special gas compositions and very careful shielding to prevent weld contamination.

Laboratory tests have shown that the high chromium plus molybdenum contents of this class of materials impart high resistance to pitting and crevice corrosion in chloride environments.* From a corrosion resistance viewpoint these materials therefore appear attractive. However, it seems likely that their availability will be linked to the future commercial viability of the production techniques used in their manufacture and field experience.

2-3 MARTENSITIC STAINLESS STEELS

2-3-1 The AISI 400 Series

Martensitic stainless steels contain more than 11.5% chromium and have an austenitic structure at elevated temperatures that can be transformed into martensite (i.e., hardened) by suitable cooling to room temperature. This requirement restricts their maximum chromium content to that defined by the region of high temperature stability of austenite, or "gamma loop," in the iron-carbon equilibrium diagram. As shown in Figure 2-10, carbon expands the gamma loop, which reaches about 18% chromium at 0.6% carbon (33). Thus the martensitic stainless steels by definition lie within the chromium range of 11.5–18%, with the lower limit being governed by corrosion resistance and the upper limit by the requirement for the alloy to convert fully to austenite on heating. To obtain mechanical properties suitable for engineering applications the steels hardened by transformation to martensite must be tempered. The resulting mechanical properties are strongly dependent on tempering temperature.

The compositions of the AISI 400 series of martensitic steels are shown in Table 2-13, and the mechanical properties in various tempered conditions are given in Table 2-14. Type 410 is the classic 12% Cr martensitic stainless steel, and is the most popular martensitic grade produced today, as indicated in Table 1-1. Applications of it are found in many wrought forms and in sand and investment castings under the Alloy Casting Institute designation CA-15. Type 403 contains lower silicon and is used in the form of forgings for turbine parts. Type 416Se is the free-machining variant containing selenium.

Type 420, with medium carbon, is the original cutlery grade, while the sulfur or selenium containing variety, type 420F, is the free-machining grade.

Types 414 and 431 contain nickel, which gives rise to somewhat higher toughness and corrosion resistance, and are used in aircraft fittings, pumps, and valves.

The high chromium-high carbon types 440A, 440B, and 440C are used

Stainless Steel '77, Climax Molybdenum Company publication.

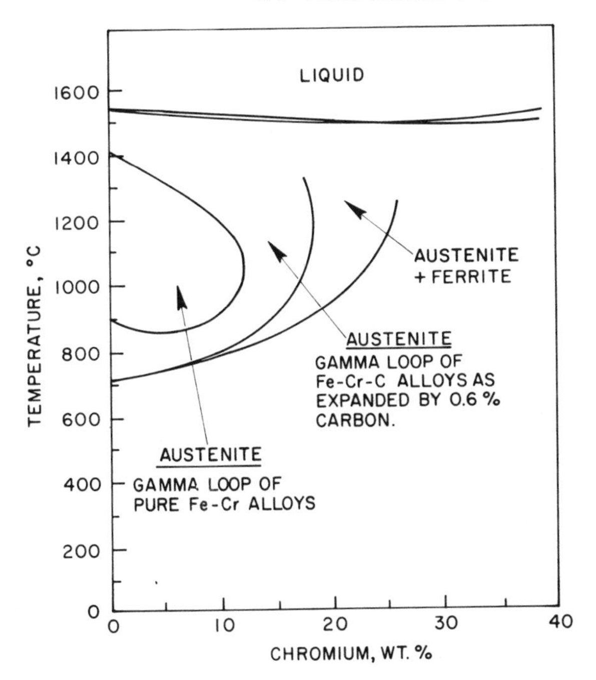

Figure 2-10. Effect of carbon addition on the gamma loop in iron-chromium alloys. (After Zapffe.)

where high hardness and wear resistance are required. Type 440F is the free-machining variety of the high carbon grades.

Type 422 contains nickel, molybdenum, tungsten, and vanadium and finds use in turbine parts and fasteners. These alloying additions enhance the development of uniform mechanical properties throughout the section of thick forgings (34). There are a number of proprietary grades of martensitic stainless steels of compositions similar to type 422 (15).

The martensitic stainless steels are generally selected for special applications which in addition to moderate corrosion resistance require some special combinations of mechanical properties (e.g., high strength with adequate toughness, corrosion plus abrasion resistance, good fatigue resistance after certain heat treatments, hardenability through thick sections, etc.). Therefore, they are largely used for turbine components, valve parts, cutlery, fasteners, and machinery parts. The high strengths that can be achieved by hardening and tempering are shown in Table 2-14.

The high carbon grades are not generally used for applications requiring welding, high ductility, a high degree of formability, or service in the

TABLE 2-13. COMPOSITIONS OF THE 400 SERIES OF MARTENSITIC STAINLESS STEELS

AISI Grade	Composition[a], (%)						
	Cr	C	Mn	Si	P	S	Other
403	11.5-13	0.15	1.0	0.5	0.040	0.030	--
410	11.5-13.5	0.15	1.0	1.0	0.040	0.030	--
414	11.5-13.5	0.15	1.0	1.0	0.040	0.030	Ni 1.25-2.50
416	12 -14	0.15	1.25	1.0	0.060	0.15[b]	Mo 0.60[d]
416Se	12 -14	0.15	1.25	1.0	0.060	0.060	Se 0.15[b]
420	12 -14	0.15[b]	1.0	1.0	0.040	0.030	--
420F	12 -14	0.38	1.25	1.0	0.060	0.15[d]	Mo 0.60[d]
422	11 -13	0.20-0.25	1.0	0.75	0.025	0.025	Ni 0.50-1.0,Mo 0.75-1.25,W 0.75-1.25,V 0.15-0.30
431	15 -17	0.20	1.0	1.0	0.040	0.030	Ni 1.25-2.50
440A	16 -18	0.60-0.75	1.0	1.0	0.040	0.030	Mo 0.75
440B	16 -18	0.75-0.95	1.0	1.0	0.040	0.030	Mo 0.75
440C	16 -18	0.95-1.20	1.0	1.0	0.040	0.030	Mo 0.75
440F[c]	17	1.0	0.4	0.4	0.04	--	S 0.08 or Se 0.18

[a]Balance iron. Single values are maximum values unless otherwise noted.
[b]Minimum.
[c]Non-standard grade, typical composition.
[d]Optional.
Source: Source Book on Stainless Steels, American Society for Metals, 1976.

TABLE 2-14. TYPICAL MECHANICAL PROPERTIES OF THE 400 SERIES OF MARTENSITIC STAINLESS STEELS

AISI Grade	Condition	Tensile Strength MPa	Yield Strength (0.2% Offset) MPa	Elongation,%	Hardness, Rockwell B or C
403,410, 416,416Se	A[b]	517	276	30	B 82
	H[c]+T at 205°C	1310	1000	15	C 41
	316°C	1241	965	15	C 39
	538°C	1000	793	20	C 31
414	A	827	655	17	C 22
	H[c]+T at 205°C	1379	1034	15	C 43
	316°C	1310	1000	15	C 41
	538°C	1000	827	20	C 34
420,420F	A	655	345	25	B 92
	H[d]+T at 316°C	1586	1345	8	C 50
422	A	793	586	22	B 98
	H[c]+T at 427°C	1627	1282	10	--
	538°C	1476	1145	13	C 42
	649°C	1000	862	14	C 32
431	A	862	655	20	C 24
	H[d]+T at 205°C	1413	1069	15	C 43
	316°C	1345	1034	15	C 41
	538°C	1034	896	18	C 34
440A	A	724	414	20	B 95
	H[d]+T at 316°C	1793	1655	5	C 51
440B	A	738	428	18	B 96
	H[d]+T at 316°C	1930	1862	3	C 55
440C,440F[a]	A	758	448	13	B 97
	H[d]+T at 316°C	1965	1896	2	C 57

[a]Non-standard grade.
[b]A = annealed.
[c]H = hardened by heating to 982°C and cooling.
[d]H = hardened by heating to 1038°C and cooling.
T = tempered at indicated temperature.
Source: Metals Handbook, American Society for Metals, Vol. 1, 1969.

temperature range 400–650°C because of temper embrittlement (35). Among corrosion resistant applications, martensitic stainless steels are used in coal handling and in mining equipment, where advantage is taken of their high abrasion resistance combined with moderate corrosion resistance.

2-3-2 Effects of Tempering

As noted in the preceding section, stainless steels hardened by transformation to martensite must be tempered to give useful engineering properties. The effect of tempering temperature on the tensile properties, impact resistance, and corrosion resistance in a saline solution of a type 420 stainless steel is shown in Figure 2-11 (35). A number of points arise from the information shown in this illustration. Tempering at temperatures in the range 450–600°C yields both poor impact resistance and poor corrosion resistance. Thus tempering temperatures used should be either above or below this critical range, giving soft or hard structures, respec-

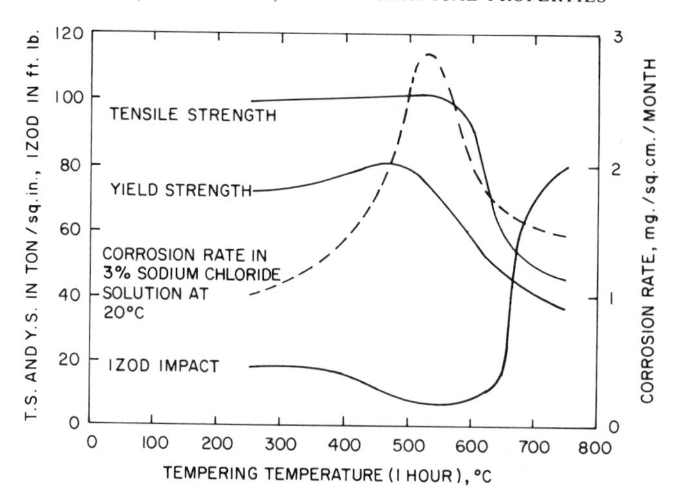

Figure 2-11. Effect of tempering on the mechanical properties and corrosion resistance of type 420 (0.22% C) stainless steel. (After Barker.)

tively. The reasons for the loss in corrosion resistance have been generally attributed to the precipitation of carbides during tempering. Studies by Truman (36) suggest that a mechanism involving chromium depletion in the vicinity of the precipitated carbides may be involved, analogous to that giving rise to sensitization in austenitic and ferritic stainless steels.

It should be noted that alloying the martensitic stainless steels with strong carbide formers, such as molybdenum, vanadium, and columbium, will stabilize the precipitated carbides in a finely dispersed form and will reduce the extent of softening due to tempering at higher temperatures. These grades, known as temper resistant types, are commercially available under various manufacturer designations.

2-4 PRECIPITATION HARDENING STAINLESS STEELS

Although the principles behind precipitation hardening of stainless steels were known in the 1930s, the first commercial material, Stainless W, did not become available until 1946. Since that time many new precipitation hardening stainless steels have been developed partly under the stimulus of the requirement of the aircraft, aerospace, and defense industries for alloys that have high strength to weight ratios combined with good corrosion resistance. While AISI type numbering exists for some of these steels in the 600 series, the 600 series numbers have not been widely used in the technical literature to describe precipitation hardening stainless steels. Description by trade designations has remained the popular practice.

As indicated in Table 2-15, the precipitation hardening stainless steels are subdivided into martensitic, semiaustenitic, and austenitic types. The martensitic types are generally supplied in the martensitic condition and precipitation hardening is achieved by a simple aging treatment of the fabricated part. The semiaustenitic types are supplied in the austenitic condition and this austenite must be transformed to martensite by special heat treatments before precipitation hardening. In the austenitic types, it is the austenite that is precipitation hardened directly.

The precipitation hardening process is thought to involve the formation of very fine intermetallics [e.g., Laves phases, $Ni_3(Al, Ti)$, carbides, and phosphides] which impede dislocation motion during deformation, giving rise to higher strength. The various types of hardening precipitates have been reviewed by Pickering (1). Prolonged aging causes these intermetallics to coarsen, enabling dislocations to bypass them during deformation, and the strength begins to decline. In this condition the material is said to be overaged. Generally, the aging treatments are designed to optimize high strength, acceptable ductility, and toughness. Precipitation hardening generally results in a slight reduction of corrosion resistance and an increase in susceptibility to hydrogen embrittlement.

The mechanical properties of the various grades in various conditions of heat treatment are shown in Tables 2-16 and 2-17, and the basic heat treatments for the various types are given in Table 2-18. In order to obtain a better understanding of the metallurgical changes underlying these heat treatments, it is instructive to refer to Figure 2-12, which shows the regions of stability of the various phases for a Fe-18% Cr-4% Ni alloy as a function of carbon content (3). The first step in the heat treatment sequence for all three classes of precipitation hardening stainless steels is solution annealing at temperatures in the single phase austenite region. The subsequent steps are determined by the stability of the austenite on cooling from this region. For all three classes, a relatively rapid cooling rate is adopted.

For the austenitic precipitation hardening steels, which are more closely represented by Figure 2-2, the nickel content is sufficiently high to ensure a fully stable austenite at room temperature. For these steels precipitation hardening is achieved by reheating the austenite to elevated temperatures at which the fine intermetallic compounds are precipitated.

The compositions of the martensitic precipitation hardening steels are adjusted to have the M_s (martensite start) temperature well above room temperature, so that a structure fully transformed to martensite is obtained at room temperature. There have been a number of attempts to establish an equation for the M_s temperature as a function of composition (1, 37, 38). An example (38) based on chromium, nickel, and carbon plus

TABLE 2-15. TYPICAL COMPOSITIONS OF THE PRECIPITATION HARDENING STAINLESS STEELS

Trade Designation	Composition[a], (%)									
	Cr	Ni	C	Mn	Si	Cu	Mo	Ti	Al	Other
				MARTENSITIC						
Stainless W[b]	16.75	6.75	0.07	0.50	0.50	--	--	0.80	0.20	--
17-4 PH	16.50	4.25	0.04	0.40	0.50	3.60	--	--	--	Cb+Ta 0.25
15-5 PH (XM-12)	15.00	4.60	0.04	0.25	0.40	3.50	--	--	--	Cb+Ta 0.35
CROLOY 16-6 PH	15.75	7.50	0.03	0.80	0.45	--	--	0.60	0.40	--
CUSTOM 450 (XM-25)	14.90	6.50	0.03	0.30	0.25	1.50	0.80	--	--	Cb+Ta 0.75
CUSTOM 455 (XM-16)	11.75	8.50	0.03	0.20	0.20	2.25	--	1.20	--	Cb+Ta 0.30
PH 13-8 Mo (XM-13)	13.00	8.00	0.04	0.05	0.05	--	2.25	--	1.00	--
ALMAR 362 (XM-9)	14.50	6.50	0.03	0.30	0.20	--	--	0.80	--	--
IN-736	10.00	10.00	0.02	0.10	0.10	--	2.00	0.20	0.30	--
				SEMI-AUSTENITIC						
17-7 PH	17.00	7.00	0.07	0.70	0.40	--	--	--	1.15	--
PH 15-7 Mo	15.00	7.00	0.07	0.70	0.40	--	2.25	--	1.15	--
AM-350	16.50	4.25	0.10	0.75	0.35	--	2.75	--	--	N 0.10
AM-355	15.50	4.25	0.13	0.85	0.35	--	2.75	--	--	N 0.12
PH 14-8 Mo[c] (XM-24)	15.50	8.75	0.05	0.10	0.10	--	2.50	--	1.35	--
				AUSTENITIC						
17-10 P	17.0	10.50	0.12	0.75	0.50	--	--	--	--	P 0.28
HNM	18.5	9.50	0.30	3.50	0.50	--	--	--	--	P 0.25
A-286	15.0	25.0	0.06	1.20	0.50	--	1.20	2.00	0.25	V 0.30

[a]Balance iron.
[b]Predominantly ferritic.
[c]Vacuum induction melted, maximum values.
Designations in parentheses are ASTM designations.
Sources: Various.

TABLE 2-16. TYPICAL MECHANICAL PROPERTIES OF THE MARTENSITIC PRECIPITATION HARDENING STAINLESS STEELS

Designation	Condition	Tensile Strength MPa[a]	Yield Strength (0.2% Offset) MPa	Elongation, %	Hardness, Rockwell C
Stainless W	A[b]	827	517	7	30
	PH[c]	1344	1241	7	46
17-4 PH	A	1034	758	10	33
	PH	1379	1227	12	44
15-5 PH	A	862	586	10	27
	PH	1379	1275	14	44
CROLOY 16-6 PH	A	924	758	16	28
	PH	1303	1275	16	40
CUSTOM 450	A	972	814	13	28
	PH	1344	1282	14	43
CUSTOM 455	A	1000	793	14	31
	PH	1724	1689	10	49
PH 13-8 Mo	A	896	586	12	28
	PH	1551	1379	13	48
ALMAR 362	A	827	724	13	25
	PH	1296	1276	15	41
IN-736	A	958	738	16	28
	PH	1310	1282	14	38

[a] 1 MPa = 145.03 psi.
[b] A = solution annealed.
[c] PH = precipitation hardened, maximum values.
Sources: Various.

nitrogen contents as variables predicts the M_s temperature (in °F) by the following relationship:

$$M_s = 2160 - 66\ (\%\ \mathrm{Cr}) - 102\ (\%\ \mathrm{Ni}) - 2620\ (\%\ \mathrm{C} + \mathrm{N}).$$

This formula was established using steels with chromium contents in the range of 10–18%, nickel in the range of 5–12.5%, and carbon plus nitrogen in the range of 0.035–0.17% (38). For a more detailed review of the effect of minor alloying elements on the M_s temperature the reader should consult the review by Pickering (1). As in the case of the austenitic steels, precipitation hardening in the martensitic steels is achieved by reheating to temperatures at which fine intermetallic compounds are precipitated. Some minor reversion of the martensite to austenite may also occur.

The class of precipitation hardening stainless steels, referred to as semiaustenitic, requires an intermediate step to transform the metastable austenite to martensite before precipitation hardening. As shown in Table 2-18, this can be achieved in two ways, either by subzero cooling, which is the more popular way, or by tempering at about 750°C to reduce the

TABLE 2-17. TYPICAL MECHANICAL PROPERTIES OF THE SEMIAUSTENITIC AND AUSTENITIC PRECIPITATION HARDENING STAINLESS STEELS

Designation	Condition	Tensile Strength MPa[a]	Yield Strength (0.2% Offset) MPa	Elongation, %	Hardness, Rockwell B or C
		SEMI-AUSTENITIC			
17-7 PH	A[b]	896	276	35	B 85
	SZC[c]+PH[d] 510°C	1620	1517	6	C 48
	T[e] 760°C,PH 566°C	1379	1275	9	C 43
PH 15-7 Mo	A	896	379	35	B 88
	SZC+PH 510°C	1655	1551	6	C 48
	T 760°C,PH 566°C	1448	1379	7	C 44
AM-350	A	1103	379	40	B 95
	T 732°C,PH 455°C	1344	1068	10.5	C 41
	SZC+PH 455°C	1379	1172	15	C 43
AM-355	A	1206	448	30	B 95
	T 732°C,PH 455°C	1344	1068	10	C 41
	SZC+PH 455°C	1517	1310	13	C 45
PH 14-8 Mo[f]	A	862	379	25	B 88
	SZC+PH 510°C	1586	1482	6	C 48
	SZC+PH 566°C	1448	1379	6	C 45
		AUSTENITIC			
17-10 P	A	613	255	70	B 82
	PH 705°C	986	676	20	C 32
HNM	A	800	386	57	B 92
	PH 732°C	1158	855	19	C 38
A-286	A	620	241	45	B 81
	PH 718°C	1006	689	25	C 34

[a] 1 MPa = 145.03 psi.
[b] A = solution annealed.
[c] SZC = sub-zero cooled to -73°C.
[d] PH = precipitation hardened at indicated temperature.
[e] T = tempered at indicated temperature.
[f] Vacuum induction melted.

Sources: Various.

Table 2-18. BASIC HEAT TREATMENTS OF PRECIPITATION HARDENING STAINLESS STEELS

Martensitic	Semi-Austenitic	Austenitic
Solution anneal in austenite region.	1. Solution anneal in austenite region.	1. Solution anneal in austenite region.
Rapidly cool to room temperature to produce a martensitic structure.	2. Rapidly cool to room temperature to produce metastable austenite.	2. Rapidly cool to room temperature to produce stable austenite.
Age at temperatures in the range 480-620°C to produce precipitation hardening.	3a. Cool to about -73°C to transform metastable austenite to martensite. OR 3b. Temper at about 760°C to reduce carbon content of austenite by precipitating carbide, then cool to room temperature to transform austenite to martensite.	3. Age at temperatures in the range 700-800°C to produce precipitation hardening.
	4. Age at temperatures in the range 450-570°C to produce precipitation hardening.	

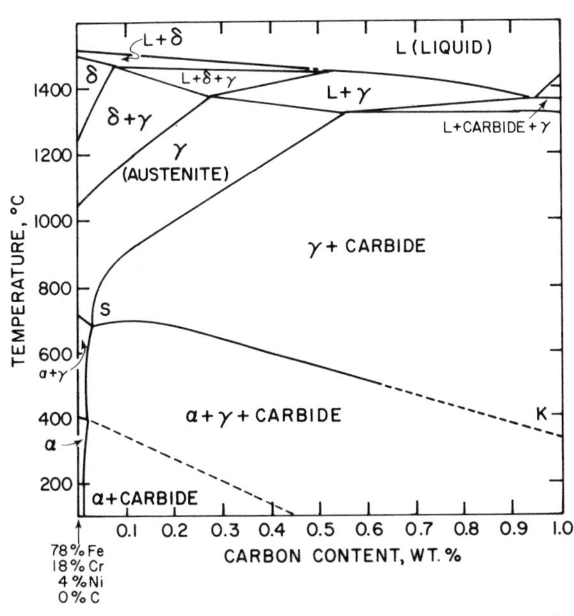

Figure 2-12. Pseudobinary phase diagram for a Fe-18% Cr-4% Ni alloy with varying carbon content. (After Krivobok.)

carbon and chromium content of the austenite by the precipitation of carbide. As can be seen from the formula for M_s, the reduction of dissolved carbon and chromium in austenite significantly raises the M_s temperature. Cold working, which also raises the M_s temperature, can also be used to accelerate the transformation of austenite to martensite.

As noted before, precipitation hardening to high strength levels introduces susceptibility to hydrogen embrittlement. Since overaging reduces this susceptibility, it is sometimes necessary to use these steels in the overaged (lower strength) conditions.

2-5 CAST STAINLESS STEELS

2-5-1 Corrosion Resistant Cast Stainless Steels

Austenitic Cast Stainless Steels. Cast stainless steels, like their wrought counterparts, can be categorized as austenitic, ferritic, martensitic, and precipitation hardening. The designations and compositions are defined by the Alloy Casting Institute (ACI). The corrosion resistant grades, used in applications where resistance to aqueous corrosion is important, have designations with the prefix C. The heat resistant grades, used in higher temperature applications where resistance to gaseous corrosion (e.g., oxidation) is necessary, have designations with the prefix H. These designations can be a little confusing when it is recognized that the nearest wrought counterpart of both CK-20 and HK is type 310. However, there are reasons for such separation. For example, the carbon content of CK-20 is 0.2% maximum, whereas that of HK is 0.6% maximum.

The compositions of the corrosion resistant cast austenitic grades are shown in Table 2-19 and mechanical properties are listed in Table 2-20. Like their wrought counterparts in the AISI 300 series, the CF grades have found wide acceptance and are the most widely used among corrosion resistant cast grades. However, while the compositions of the wrought stainless steels are usually carefully balanced to avoid delta ferrite for improved workability, the CF cast grades can contain significant amounts of delta ferrite. For example, the molybdenum containing CF-8M (type 316) usually contains 5–20% delta ferrite distributed throughout the austenite matrix in the form of discontinuous pools (39). This grade is used in moderate acid and chloride environments where the molybdenum content imparts pitting resistance.

As in the case of the higher carbon wrought grades, the cast CF grades can also exhibit poor resistance to intergranular corrosion due to chromium depletion in the vicinity of precipitated carbides. This can be prevented by solution annealing the casting at temperatures well in the austenitic range (e.g., 1100°C as shown in Figure 2-2). Even the low carbon grades, CF-3 and CF-3M, in the as-cast condition can contain

TABLE 2-19. COMPOSITIONS OF CORROSION RESISTANT CAST AUSTENITIC STAINLESS STEELS

ACI Designation	Nearest Wrought Comparative	Composition[a], (%) Cr	Ni	C	Mn	Si	Other
CF- 3	304L	18-21	8-11	0.03	1.5	2.0	--
CF- 8	304	18-21	8-11	0.08	1.5	2.0	--
CF-20	302	18-21	8-11	0.20	1.5	2.0	--
CF-3M	316L	18-21	9-12	0.03	1.5	1.5	Mo 2.0-3.0
CF-8M	316	18-21	9-12	0.08	1.5	1.5	Mo 2.0-3.0
CF-8C	347	18-21	9-12	0.08	1.5	2.0	Cb[b]
CF-16F	303	18-21	9-12	0.16	1.5	2.0	Mo 1.5,Se 0.2-0.35
CG-8M	317	18-21	8-11	0.08	1.5	1.5	Mo 3.0 (min.)
CH-20	309	22-26	12-15	0.20	1.5	2.0	--
CK-20	310	23-27	19-22	0.20	1.5	2.0	--
CN-7M	20Cb-3	18-22	21-31	0.07	1.5	[c]	Cu 3.0,Mo 2.0

[a]Balance iron. Single values are maximum values unless otherwise noted.
[b]Cb 8xC min., 1.0% max., or Cb+Ta 10xC min., 1.35% max.
[c]Variable.
Source: Metals Handbook, American Society for Metals, 1969.

TABLE 2-20. TYPICAL MECHANICAL PROPERTIES OF CORROSION RESISTANT CAST AUSTENITIC STAINLESS STEELS

ACI Designation	Nearest Wrought Comparative	Tensile Strength MPa[a]	Yield Strength (0.2% Offset) MPa	Elongation, %	Hardness, Brinell
CF- 3	304L	531	248	60	140
CF- 8	304	531	255	55	140
CF-20	302	531	248	50	163
CF-3M	316L	531	276	50	163
CF-8M	316	552	290	50	163
CF-8C	347	531	262	39	149
CF-16F	303	531	276	52	150
CG-8M	317	572	331	45	170
CH-20	309	607	345	38	190
CK-20	310	524	262	37	144
CN-7M	20Cb-3	476	220	48	130

[a]1 MPa = 145.03 psi.
Sources: Various.

TABLE 2-21. COMPOSITIONS OF CORROSION RESISTANT CAST HIGHER ALLOYS

ASTM Designation	Nearest Wrought Comparative	Typical Composition, (%)				
		Cr	Ni	C (max.)	Mo	Other
IN-862[a]	AL-6X	21	24	0.07	5.0	Fe,Bal.
CW-12M-1[b]	HASTELLOY Alloy C-276	16	51	0.12	17	V,W,Fe,Co
CW-12M-2[c]	--	18	58	0.07	18	--
Alloy 625[a]	INCONEL Alloy 625	21.5	60	0.03	9	Cb
CY-40	INCONEL Alloy 600	15	74	0.40	--	Fe

[a]Designation of International Nickel Company, Inc.
[b]Includes proprietary alloy ILLIUM W-1.
[c]Includes proprietary alloys CHLORIMET 3 and ILLIUM W-2.
Sources: Various.

TABLE 2-22. COMPOSITIONS OF CORROSION RESISTANT CAST MARTENSITIC, FERRITIC, AND PRECIPITATION HARDENING STAINLESS STEELS

ACI Designation	Nearest Wrought Comparative	Composition[a], (%)					
		Cr	Ni	C	Mn	Si	Other
		MARTENSITIC					
CA-15	410	11.5-14	1.0	0.15	1.0	1.5	Mo 0.5[b]
CA-40	420	11.5-14	1.0	0.40	1.0	1.5	Mo 0.5[b]
CA-6NM	--	11.5-14	4.0	0.06	1.0	1.5	Mo 0.7
CA-6N	--	11.5-14	7.0	0.06	1.0	1.5	Al,Ti
		FERRITIC					
CB-30	442	18 -22	2.0	0.30	1.0	1.0	--
CC-50	446	26 -30	4.0	0.50	1.0	1.0	--
CD-4MCu[c,d]	329	26	5.0	0.04	1.0	1.0	Cu 3.0, Mo 2.0
		PRECIPITATION HARDENING					
CB-7Cu-1[c]	17-4 PH	16	4.0	0.05	--	--	Cu 3.0
CB-7Cu-2[c]	15-5 PH	15	5.0	0.05	--	--	Cu 3.0,Cb

[a]Balance iron. Single values are maximum values unless otherwise noted.
[b]Not intentionally added.
[c]Typical compositions.
[d]Duplex, predominantly ferritic.
Sources: International Nickel Company, Inc.
Metals Handbook, American Society for Metals, 1969.

grain boundary carbides and should be solution annealed before welding. Reannealing is generally not required after welding. A columbium stabilized grade CF-8C (type 347) is also available for welded applications.

A free-machining grade containing selenium, CF-16F, is also available. However, this grade is not widely used since the cast CF-8 (type 304) is not particularly difficult to machine by modern techniques (39).

As in the case of wrought austenitic alloys, there are instances where the corrosion resistance of the cast austenitic steels is not adequate to withstand severe environments. In such instances the search for more corrosion resistant castings must be extended to higher alloys containing less than 50% iron. Some currently available higher alloy castings are shown in Table 2-21. These are the cast counterparts of some of the proprietary wrought higher alloys identified in Tables 2-7 and 2-8, and for information regarding their properties the reader should consult the supplier literature.

Martensitic, Ferritic, and Precipitation Hardening Cast Stainless Steels. The compositions of these cast steels are shown in Table 2-22 and their mechanical properties are given in Table 2-23. As in the case of their

TABLE 2-23. TYPICAL MECHANICAL PROPERTIES OF CORROSION RESISTANT CAST MARTENSITIC, FERRITIC, AND PRECIPITATION HARDENING STAINLESS STEELS[a]

ACI signation	Nearest Wrought Comparative	Tensile Strength MPa[b]	Yield Strength (0.2% Offset) MPa	Elongation,%	Hardness, Brinell
		MARTENSITIC			
CA-15	410	931	793	17	390
CA-40	420	1034	862	10	470
CA-6NM	--	827	689	24	300
CA-6N	--	965	931	15	--
		FERRITIC			
CB-30	442	655	414	15	195
CC-50	446	669	448	18	210
D-4MCu[c]	329	745	558	25	260
		PRECIPITATION HARDENING			
B-7Cu-1	17-4 PH	--	1076	3	418
B-7Cu-2	15-5 PH	1241	1034	6	--

fter heat treatment for optimum properties.
MPa = 145.03 psi.
an be precipitation hardened. Properties noted here are for the olution annealed condition.
urces: Various.

TABLE 2-24. TYPICAL COMPOSITIONS AND CREEP RATES OF HEAT RESISTANT CAST STAINLESS STEELS

ACI Designation	Wrought Comparative	Composition[a], (%)				Creep Rate	
		Cr	Ni	C	Si (max.)	Temp., °C	Stress to Give Creep of 0.0001%/Hr., MPa[c]
HA	--	9	--	0.20[b]	1.0	649	21.4
HC	446	28	4[b]	0.50[b]	2.0	871	5.2
HD	--	28	5.5	0.50[b]	2.0	982	6.2
HE	--	28	9.5	0.20-0.50	2.0	982	9.7
HF	302B	21	10.5	0.20-0.40	2.0	871	26.9
HH	309	26	12.5	0.20-0.50	2.0	982	14.5
HI	--	28	16	0.20-0.50	2.0	982	13.1
HK	310	26	20	0.20-0.60	2.0	1038	9.7
HL	--	30	20	0.20-0.60	2.0	982	15.2
HN	--	21	25	0.20-0.50	2.0	1038	11.0
HP	--	26	35	0.35-0.75	2.5	982	19.3
HT	330	17	35	0.35-0.75	2.5	982	13.8
HU	--	19	39	0.35-0.75	2.5	982	15.2
HW	--	12	60	0.35-0.75	2.5	982	9.6
HX	--	17	66	0.35-0.75	2.5	982	11.0

[a]Balance iron.
[b]Maximum.
[c]1 MPa = 145.03 psi.
Source: International Nickel Company, Inc.

wrought counterparts, the cast martensitic alloys can be hardened and tempered to yield a wide range of mechanical properties; they are used where moderate corrosion resistance together with high abrasion resistance are required.

The ferritic cast alloys cannot be hardened and use is made of their high chromium content which imparts corrosion resistance in oxidizing acids. The molybdenum containing duplex alloy CD-4MCu has good pitting and stress corrosion resistance in chloride environments.

The precipitation hardening alloys combine the highest strengths attainable in cast stainless steels with good toughness and moderate corrosion resistance.

2-5-2 Heat Resistant Cast Stainless Steels

The compositions of the ACI grades of heat resistant stainless steel castings are given in Table 2-24, together with creep properties at certain elevated temperatures. The iron-chromium martensitic and ferritic variants HA and HC are used chiefly for resistance to oxidation since they have comparatively low strength at elevated temperatures and can become embrittled in certain temperature ranges (40). The iron-chromium-nickel variants HD, HE, HF, HH, HI, HK, and HL have greater high temperature strength than the iron-chromium grades as well as resistance to oxidizing and sulfur containing reducing gases. The iron-nickel-chromium variants HN, HP, HT, HU, HW, and HX have high strength at elevated temperatures and resistance to carburizing or nitriding atmospheres. At the higher temperature-higher strength end the cast stainless steels merge with cast nickel base and cobalt base alloys used as investment castings in jet engine applications.

REFERENCES

1. F. B. Pickering, *Int. Met. Rev.*, p. 227, December 1976.
2. A. Hanson and J. G. Parr, *The Engineer's Guide to Steel*, Addison-Wesley, Reading, Mass., 1965, p. 265.
3. V. N. Krivobok, in *The Book of Stainless Steels*, E. E. Thum, Ed., The American Society for Steel Treating, Cleveland, Ohio, 1933.
4. C. Husen and C. H. Samans, *Chem. Eng.*, p. 178, January 27, 1969.
5. B. Strauss, H. Schottky, and J. Hinnuber, *Z. Anorg. Chem.*, Vol. 188, p. 309, 1930.
6. E. C. Bain, R. H. Aborn, and J. B. Rutheford, *Trans. Am. Soc. Steel Treat.*, Vol. 21, p. 481, 1933.
7. L. R. Honnaker, *Chem. Eng. Prog.*, Vol. 54, p. 79, 1958.
8. M. Henthorne, "Intergranular Corrosion in Iron and Nickel Base Alloys," *Localized Corrosion—Cause of Metal Failure*, ASTM-STP 516, American Society for Testing and Materials, 1972, p. 66.
9. R. P. Reed, *Acta Met.*, Vol. 10, p. 865, 1962.
10. V. N. Krivobok and A. M. Talbot, *Proc. ASTM*, Vol. 50, p. 859, 1950.
11. M. L. Holzworth and M. R. Louthan, *Corrosion*, Vol. 24, No. 4, p. 110, 1968.

12. S. S. Birley and D. Tromans, *Corrosion,* Vol. 27, No. 2, p. 63, 1971.
13. M. Henthorne, *Sulfide Inclusions in Steel*, American Society for Metals, Metals Park, Ohio, 1975, p. 445.
14. H. E. McGannon, Ed., *The Making, Shaping and Treating of Steel,* United States Steel Corporation, Pittsburgh, Pa., 1964, p. 1116.
15. *Source Book on Stainless Steels,* American Society for Metals, Metals Park, Ohio, 1976, p. 392.
16. J. A. Chivinsky, *Met. Prog.,* p. 55, February 1972.
17. R. C. Gibson, H. W. Hayden, and J. H. Brophy, *Trans. ASM,* Vol. 61, p. 85, 1968.
18. V. N. Krivobok, *Trans. ASM,* Vol. 23, p. 1, 1935.
19. *Metals Handbook,* 8th ed., *Properties and Selection of Metals,* Vol. 1, American Society for Metals, Metals Park, Ohio, 1969, p. 421.
20. R. A. Lula, *Met. Prog.,* p. 24, July 1976.
21. R. M. Fisher, E. J. Dulis, and K. G. Carroll, *Trans. AIME,* Vol. 197, p. 690, 1953.
22. R. O. Williams, *Trans. AIME-TMS,* Vol. 212, p. 497, 1958.
23. P. J. Grobner, *Met. Trans.,* Vol. 4, p. 251, 1973.
24. J. J. Demo, *Structure, Constitution, and General Characteristics of Wrought Ferritic Stainless Steels,* ASTM STP-619, American Society for Testing and Materials, 1977.
25. M. E. Nicholson, C. H. Samans, and F. J. Shortsleeve, *Trans. ASM,* Vol. 44, p. 601, 1952.
26. E. C. Hoxie, in *Pressure Vessels and Piping: Decade of Progress,* Vol. 3, The American Society of Mechanical Engineers, New York, 1977.
27. Reference 19, p. 427.
28. A. Bäumel, *Arch. Eisenhaüttenwes.,* Vol. 34, p. 135, 1963.
29. E. Baerlecken, W. A. Fisher, and K. Lorenz, *Stähl Eisen,* Vol. 81, p. 768, 1961.
30. J. J. Demo, *Corrosion,* Vol. 27, p. 531, 1971.
31. A. P. Bond and E. A. Lizlovs, *J. Electrochem. Soc.,* Vol. 116, p. 1305, 1969.
32. R. J. Hodges, *Corrosion,* Vol. 27, p. 119, 1971.
33. C. A. Zapffe, *Stainless Steels,* The American Society for Metals, Cleveland, Ohio, 1949, p. 106.
34. Reference 19, pp. 411 and 415.
35. R. Barker, *Metallurgia,* p. 49, August 1967.
36. J. E. Truman, *Br. Corros. J.,* Vol. 11, No. 2, p. 92, 1976.
37. G. H. Eichelman and F. C. Hull, *Trans. ASM,* Vol. 45, p. 77, 1953.
38. F. C. Monkman, F. B. Cuff, and N. J. Grant, *Met. Prog.,* p. 94, April 1957.
39. Reference 19, p. 433.
40. Reference 19, p. 446.

3

ELECTROCHEMISTRY

3-1 MIXED POTENTIAL THEORY

The last two decades have seen a considerable increase in the use of electrochemical techniques to study the corrosion behavior of stainless steels. From a technological viewpoint this increase has been stimulated primarily by the desire to develop a capability to predict by short term laboratory tests the behavior of stainless steels in industrial and natural environments. This increase has also been aided by the development of commercially available potentiostats, as well as other instruments for measuring corrosion rates from electrochemical behavior. Electrochemical techniques are now so often used in corrosion testing that it is very desirable that anyone confronted with the topic of corrosion of stainless steels become familiar with the basic essentials, benefits, and limitations of electrochemical evaluations.

The theoretical basis for electrochemical corrosion testing is derived from the mixed potential theory, the formulation of which in its modern form is usually attributed to Wagner and Traud (1). In essence this theory separates the oxidation and reduction reactions of corrosion and postulates that the total rates of all the oxidation reactions equal the total rates of all the reduction reactions on a corroding surface.

Oxidation reactions, referred to as anodic, because they occur at the anodic sites on a corroding metal or at the anode in an electrochemical cell, can be represented by the general reaction

$$M \rightarrow M^{+n} + ne.$$

This is the generalized corrosion reaction that removes the metal atom by oxidizing it to its ion. In this reaction the number of electrons produced equals the valence of the metal ion produced. The mixed potential theory proposes that all the electrons generated by the anodic reactions are consumed by corresponding reduction reactions. Reduction reactions are known as cathodic, because they occur at the cathodic sites of a corroding metal or at the cathode of an electrochemical cell. The more common cathodic (electron consuming) reactions encountered in aqueous corrosion are as follows:

1. Reduction of hydrogen ions—$2H^+ + 2e \rightarrow H_2$.
2. Oxygen reduction (acid solutions)—$O_2 + 4H^+ + 4e \rightarrow 2H_2O$.
3. Oxygen reduction (basic or neutral solutions)—$O_2 + 2H_2O + 4e \rightarrow 4OH^-$.
4. Metal ion reduction—$M^{+n} + e \rightarrow M^{+(n-1)}$.
5. Metal deposition (plating)—$M^{+n} + ne \rightarrow M$.

During corrosion more than one anodic and more than one cathodic reaction may be operative. Consider, for example, the corrosion of a stainless steel in a hydrochloric acid solution contaminated with ferric ions. By means of the anodic reaction, $M \rightarrow M^{+n} + ne$, all the component elements (i.e., iron, chromium, etc.) of the alloy go into the solution as their respective ions. The electrons produced by these anodic reactions will be consumed by the cathodic reactions 1 and 4 above, with reaction 4 in this instance being represented by $Fe^{+3} + e \rightarrow Fe^{+2}$. Removing one of the available cathodic reactions (e.g., reaction 4, by the removal of the ferric ions) will reduce the corrosion rate. Thus the observation that hydrochloric acid contaminated with ferric ions is more corrosive than pure hydrochloric acid can be easily explained in terms of the mixed potential theory of corrosion.

The mixed potential itself, which is commonly referred to as the corrosion potential, denoted by the symbol E_{corr}, is the potential at which the total rates of all the anodic reactions are equal to the total rates of all the cathodic reactions. The current density at E_{corr} is called the corrosion current density, i_{corr}, and is a measure of the corrosion rate.*

3-2 MEASUREMENT OF CORROSION RATE BY EXTRAPOLATION

The corrosion current density, i_{corr}, cannot be measured directly, since the current involved is one that flows between numerous microscopic anodic and cathodic sites on the surface of the corroding metal. However, in some cases including that of a stainless steel freely corroding in a nonoxidizing acid, it is possible to measure this current indirectly with the aid of a counter electrode (usually platinum) and electronic equipment. While there are several techniques to do this, a convenient modern

*General corrosion (active dissolution) rates of stainless steels in acids or alkalies are often measured by the weight loss techniques described in Chapter 9.

practice is to employ a potentiostat in conjunction with a reference electrode (often a saturated calomel electrode). Essentially, the potentiostat is an instrument that applies a current to a specimen which enables the potential to be controlled in a desired way. The reference electrode in the circuit provides a known standard potential enabling comparisons to be made from test to test.

One way to measure i_{corr} is by extrapolating certain linear segments of measured potential-current density curves. The principles behind the extrapolation technique can be understood from the following procedure. In this procedure the specimen is initially made to act as a cathode in the electrochemical cell containing the corrodent (electrolyte). The cathodic potential-current density curve is measured over the potential range defined by E_{corr} and some potential active to E_{corr}, for example, E_C in Figure 3-1, and is shown schematically by the continuous line. The curve can be established either by a stepwise or continuous changing of potential. At E_{corr} the measured cathodic current density approaches zero.

Increasing the applied potential in the noble direction away from E_{corr} now makes the specimen behave as an anode. The anodic current increases with increasing noble potential, giving rise to the measured anodic curve, shown in Figure 3-1.

In theory both the cathodic and anodic potential-current density variations should be linear, intersecting at a point defined by E_{corr}/i_{corr}. However, the measured curves deviate from linearity on approaching E_{corr} (Figure 3-1). Without going into extended discussion, this deviation from

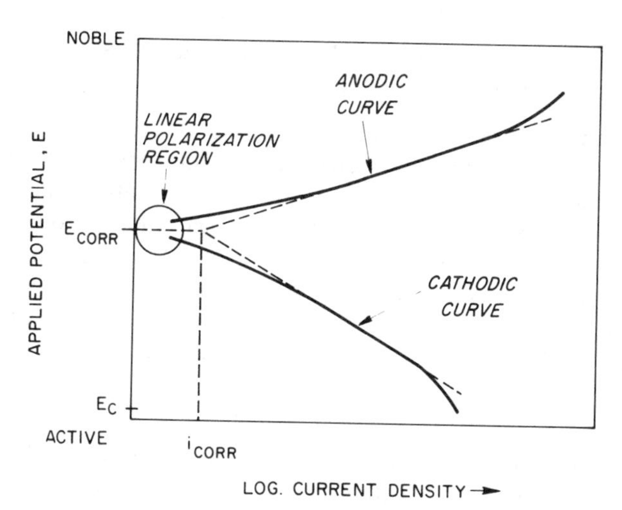

Figure 3-1. Measurement of i_{corr} by extrapolation.

linearity is thought to be a consequence of both anode and cathode sites developing on the specimen surface as the applied currents become vanishingly small. Nevertheless, both the measured cathodic and anodic potential-current density curves contain linear segments, sometimes referred to as Tafel regions, named after Tafel (2). Extrapolating the linear segments of either the measured anodic or cathodic curves and intersecting these extrapolated lines with a line drawn from E_{corr} parallel to the current density axis will yield i_{corr}, as shown in Figure 3-1. Extrapolations from the linear segment of the cathodic curve are generally preferred since the latter is often easier to measure experimentally (3, 4). This extrapolation procedure dates back to the studies of Wagner and Traud (1).

There are several practical procedures described in the corrosion literature for measuring i_{corr} by the extrapolation technique, including both potentiostatic and galvanostatic procedures (3–6). A recent potentiostatic procedure (3), which has been used to measure i_{corr} of stainless steels in nonoxidizing acids, employs several cathodic to anodic potential sweeps to establish the linear cathodic region, in conjunction with a fast potential sweeping rate made available by the latest commercial potentiostats.

The corrosion current density, i_{corr}, can be converted to corrosion rate by the relationship

$$R_{mpy} = 0.13\, i_{corr} \frac{e}{\rho}$$

where R_{mpy} = corrosion rate (mils/year)
i_{corr} = corrosion current density ($\mu A/cm^2$)
e = equivalent weight (chemical) of the metal
ρ = density of the metal (g/cm^3).

This equation describes the corrosion rate of a pure metal which has a certain density and equivalent weight. Since stainless steels comprise a number of major alloying elements of differing densities and equivalent weights, a computation must be made of the partial contributions of the various alloying elements. Such computations have been made for a number of stainless steels and higher alloys and are given in Table 3-1 (3). These conversion factors, K, for each alloy listed in Table 3-1, multiplied by the measured i_{corr} yield the corrosion rate; that is

$$R_{mpy} = K i_{corr}.$$

The extrapolation technique for measuring i_{corr} is dependent on the ability to identify the linear (Tafel) region. Corrodents in which more than one reduction reaction is operative or in which the rate of arrival of the

TABLE 3-1. CONVERSION FACTORS FOR CORROSION RATE CALCULATION

Alloy	Conversion Factor, K
AISI Type 304	0.53
AISI Type 316	0.55
CARPENTER 20Cb-3	0.52
INCOLOY alloy 800	0.53
INCOLOY alloy 825	0.43
INCONEL alloy 600	0.48
HASTELLOY alloy C	0.67
INCONEL alloy 625	0.58

reducible species (e.g., hydrogen ions) at the cathode surface begins to determine the reduction rate (known as concentration polarization) exhibit less distinct linear regions, making the extrapolation less certain. These disadvantages can largely be overcome by the linear (resistance) polarization technique.

3-3 MEASUREMENT OF CORROSION RATE BY LINEAR (RESISTANCE) POLARIZATION

The value of i_{corr} can also be measured by another technique, generally known as linear polarization. This technique is based on the theoretical and practical demonstration that at potentials very close to E_{corr}, ± 10 mV, the slope of the potential/applied current curve is approximately linear, as shown in Figure 3-2. This slope, $\Delta E/\Delta i$, has the units of resistance; hence the technique is sometimes called the polarization resistance method. The region defined by $E_{corr} \pm 10$ mV, which is the circled region in Figure 3-1, is difficult to study by potentiostatic techniques. However, a number of commercial instruments are readily available for linear polarization measurements that incorporate a probe containing the test, reference and auxiliary electrodes, a current source, ammeter, voltmeter, and a display panel. The principles behind the various types of presently available commercial linear polarization equipment have been described in a recent publication (4). These types of equipment are beginning to find increasing use in monitoring corrosion rates in industrial plants.

Accepting the linearity of the slope $\Delta E/\Delta i$, it has been shown (7, 8) that i_{corr} is related to the inverse of this slope by the equation

$$i_{corr} = \left[\frac{\beta_a \beta_c}{2.3(\beta_a + \beta_c)}\right] \frac{\Delta i}{\Delta E}$$

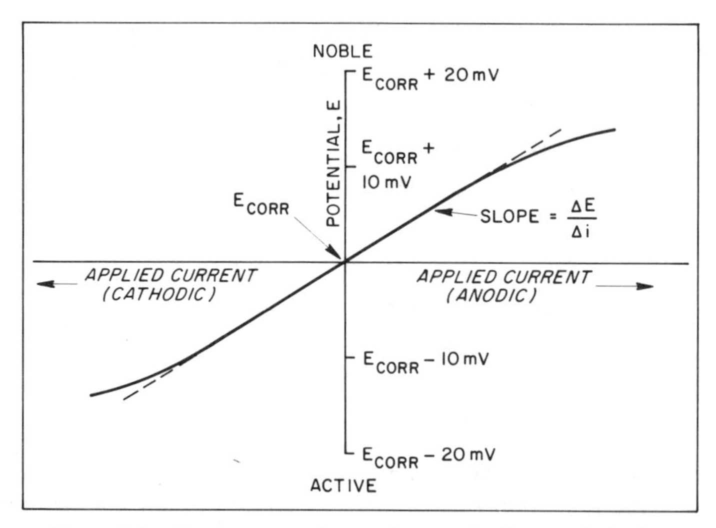

Figure 3-2. Measurement of corrosion rate by linear polarization.

where β_a and β_c are the anodic and cathodic Tafel slopes, respectively. It is generally accepted that the quantity $\beta_a\beta_c/2.3(\beta_a + \beta_c)$ is a constant (C). Hence, i_{corr} is given by the relationship

$$i_{corr} = C\left(\frac{\Delta i}{\Delta E}\right)$$

with C being measured experimentally or estimated (9, 10).

In the last decade a number of papers have appeared that have questioned the validity of the concepts behind the linear polarization technique for measuring i_{corr}. A review and discussion of this topic is contained in the publication by Shreir (9). However, it should be noted that in the earlier work (8) it was shown that corrosion rates determined by linear polarization are in good agreement with corrosion rates obtained by weight loss methods. A more recent appraisal employing mathematical and graphic methods has led to the conclusion that the various assumptions of linearity are sufficient for the technique to be valid in many practical corrosion systems (11). Accordingly, it must be accepted that linear polarization is a useful, if approximate, electrochemical technique for measuring corrosion rates.

3-4 THE ACTIVE-PASSIVE TRANSITION AND PASSIVITY

So far the discussion has centered on the application of electrochemical techniques to measure the corrosion rates of stainless steels corroding in aggressive (e.g., reducing acid) environments. While stainless steels are sometimes used under such conditions, it is usually the exception rather than the rule. It is usually the economic goal to select a stainless steel, or higher alloy, capable of withstanding a given environment for the lifetime of the equipment in question. The state in which a stainless steel exhibits a very low corrosion rate is known as passivity.* It is probably in the area of defining passivity, and its limits, that the mixed potential theory and the electrochemical approach to defining corrosion has made the greatest technological impact. A direct consequence of the application of mixed potential theory has been the development of a corrosion control technique known as anodic protection. While the same cannot be claimed for cathodic protection, the concept of which goes back some 150 years, the latter has also benefited in terms of economic refinement and improved understanding.

A convenient way to obtain an understanding of the passivity of stainless steels is by considering the potential-current density diagram, generally known as a polarization curve. Continuing the experiment initiated in Figure 3-1, namely, increasing the applied potential in the noble direction from E_{corr} and recording the resulting current, ultimately yields a somewhat unexpected potential current-density variation, as shown in Figure 3-3. First, the measured current ceases to increase with applied potential, and at a potential usually called E_{pp} (the primary passivation potential) it begins to decrease. The beginning of this decrease is known as the active-passive transition. Above this potential, the current drops to a very low value, i_{pass}, (the passive current density) and remains at a low value (although in practice not necessarily constant) over a wide range of potentials. The potential range over which the current remains at a low value is termed the passive potential range, and it is this range which defines passivity for a given stainless steel/environment combination. Corrosion rates in this passive range are usually very low.

On continuing to increase the applied potential in the noble direction, another potential will be reached at which the measured current will again begin to increase. The potential at which this current increase takes place is critically dependent on the corrodent. For example, in a chloride free aqueous sulfuric acid solution this potential represents the onset of the evolution of gaseous oxygen† by the electrolysis of water, and is known as E_t, the potential representing the onset of transpassive behavior. However, in chloride containing solutions the current "breaks" at somewhat

*For a recent discussion of this subject the reader should consult *Passivity of Metals*, The Electrochemical Society, Princeton, N. J., 1978.

†Some metal dissolution is also taking place.

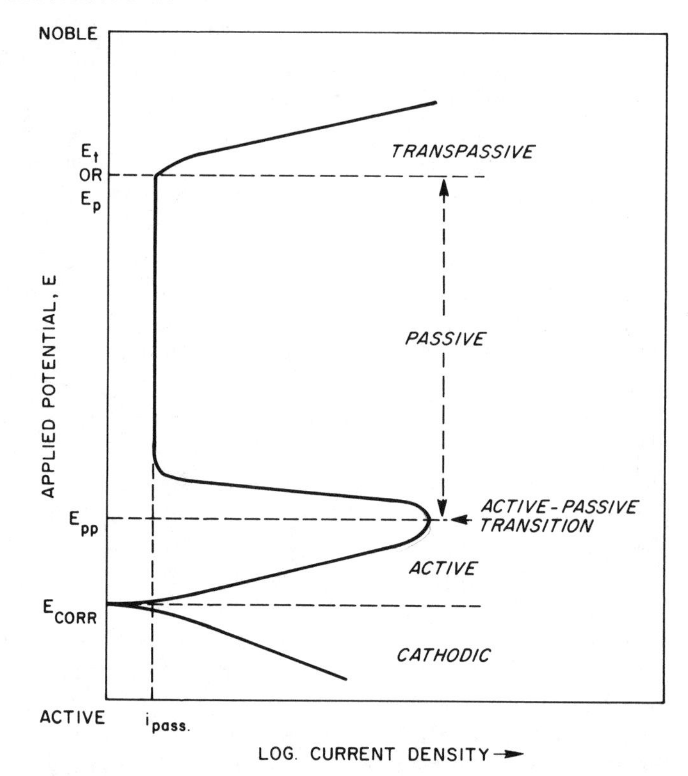

Figure 3-3. A schematic polarization curve for a stainless steel in a sulfuric acid solution.

lower (more active) potentials, as shown in Figure 3-4. This "breakthrough" is accompanied by the formation of corrosion pits on the test specimen surface, and is usually known as the pitting potential, E_p. Since pitting can perforate and destroy industrial equipment, the potential E_p represents a limiting potential which should not be exceeded. As many industrial environments are contaminated with or contain chloride ions, the passive potential range can be practically defined as the range between E_{pp} and E_p, although in some cases it is defined by the range between E_{pp} and E_t.

Clearly, it is to the benefit of an operator of stainless steel industrial equipment to ensure that the material remains in the passive potential range during operation. This can be accomplished by selecting the right material/environment combination. Test procedures, such as establishing a corrosion behavior diagram (CBD) for various material/environment combinations have been used (3), and are finding increasing application in

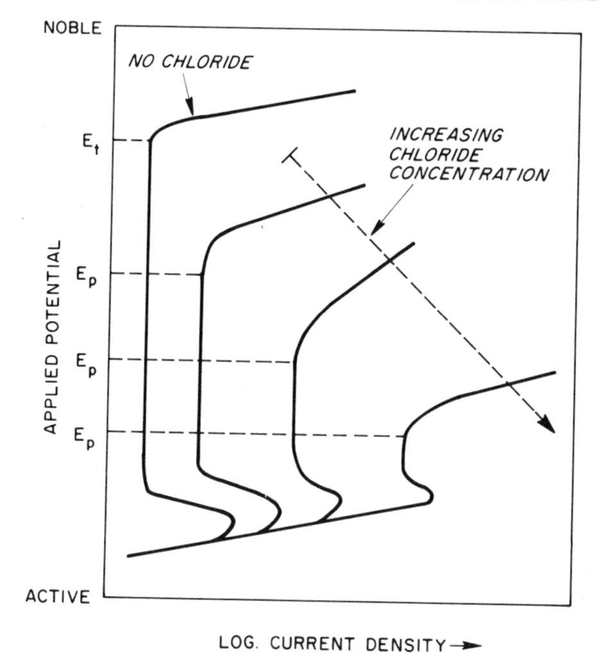

Figure 3-4. Typical polarization curves for a stainless steel in a sulfuric acid solution containing increasing amounts of chloride.

material selection. Alternatively the designer can select a material, often less expensive initially, which will not be in the passive state during operation, and artificially protect his equipment by the corrosion prevention techniques of anodic protection or cathodic protection. In anodic protection the potential of the engineering material is brought into the passive range by the application of an external current. In cathodic protection, not often used for stainless steels, the equipment is made to be a cathode by electrically connecting it to a sacrificial anode or by applying a current. These two techniques are discussed in the next section.

3-5 ANODIC AND CATHODIC PROTECTION

The development of anodic protection to prevent the corrosion of stainless steels in sulfuric acid dates back to 1954 and is attributed to Edeleanu (12, 13), who made use of polarization curves to determine the optimum conditions for protection and demonstrated the practicality of the technique in a pilot plant. However, the technique was adopted by industry only some 15 years ago. It is restricted to material/environment combi-

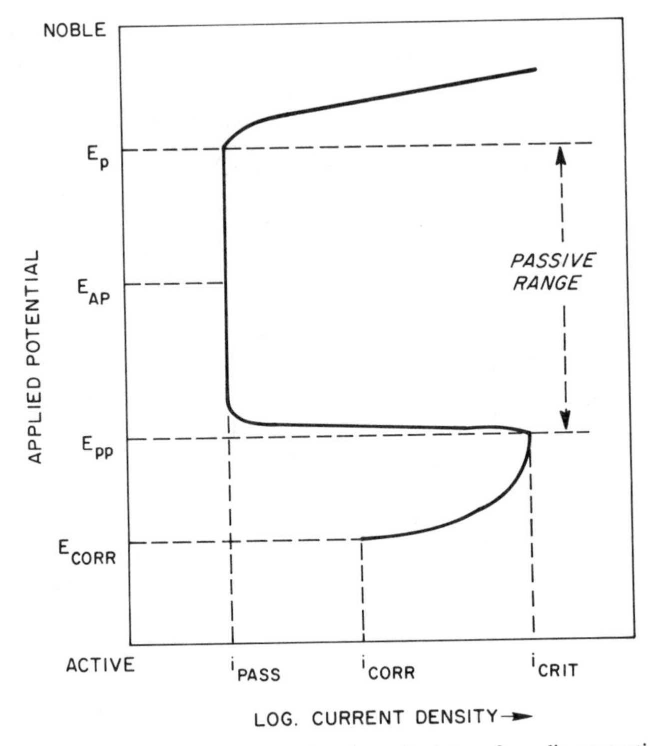

Figure 3-5. Polarization curve showing principles of anodic protection.

nations that exhibit active-passive transitions and has been used mainly to protect stainless steels in sulfuric acid environments.

The principles behind anodic protection are shown in Figure 3-5. Assume that a given stainless steel structure (e.g., holding tank, shell and tube heat exchanger, etc.) exhibits an unacceptably high corrosion rate, i_{corr}, at its natural corrosion potential E_{corr}. In anodic protection the aim is to reduce the corrosion rate to a value indicated by i_{pass}. (Note that the current scale in Figure 3-5 is logarithmic.) To do this requires the raising of the corrosion potential to a value within the passive range, preferably in the middle of it, at E_{AP}. This is achieved by applying a current density, i_{crit}. Once this has been accomplished, it takes a much smaller current density to maintain the potential at E_{AP}. The current consumed to maintain passivity, once it has been achieved, may be two orders of magnitude less than that needed to achieve it.

As is evident from Figure 3-4, the width of the passive range and the potentials defining it are sensitive to contaminants, such as chloride ions.

It is therefore very important to determine E_{AP} in the actual process acid of interest at maximum and minimum levels of contaminants and process temperatures. Furthermore, in process acids there may be contaminants present which may undergo electron releasing (anodic) reactions unrelated to the achievement of passivity and hence obscure the active-passive behavior of the stainless steel. Recent studies (14) have shown that under such conditions it may be possible to estimate E_{AP} by taking the midpoint of a low current region established by a forward plus reverse polarization scan. This practice is shown schematically in Figure 3-6, with the potential scanning direction indicated by the sequence ABCDEF. The segment DE denotes the low current region, with E_{AP} being taken as its midpoint. The anodic current increase at B is not the active-passive transition but denotes some unspecified reaction on the electrode surface. If B was mistakenly assumed to be the active-passive transition and the anodic protection potential estimated as the apparent "E_{AP}" in Figure 3-6, high corrosion rates would be obtained if anodic protection were attempted at this potential. While the reverse scanning technique requires further evaluation in terms of broad application as a method for determining E_{AP}, it appears to be a promising technique in cases where extraneous reactions may occur on the anode.

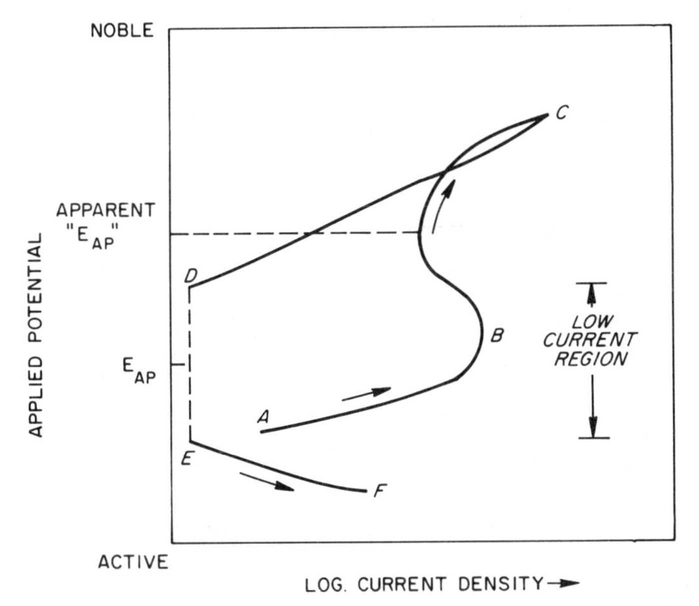

Figure 3-6. Estimate of anodic protection potential from low current region observed on reverse polarization. Anodic peak at B is not the active-passive transition.

Cathodic protection, as noted earlier, is not an outgrowth of the mixed potential theory, but dates back, at least conceptually, to 1824 when Sir Humphry Davy in England described how zinc anodes could be used to prevent the corrosion of the copper sheathing of the wooden hulls of ships. It has been pointed out (15) that it was unfortunate that cathodic protection of these copper sheaths not only inhibited corrosion but also increased marine fouling rate. An upsurge in interest in cathodic protection came later with the need for protection of undergound pipes.

Figure 3-7 shows that the principles behind cathodic protection can be conveniently described in terms of the potential-current density relationships used in discussing the electrochemistry of stainless steels. Figure 3-7 *a* describes the corrosion rate of a metal, i_{corr}, at its corrosion potential E_{corr}. To obtain complete cathodic protection the potential of the metal

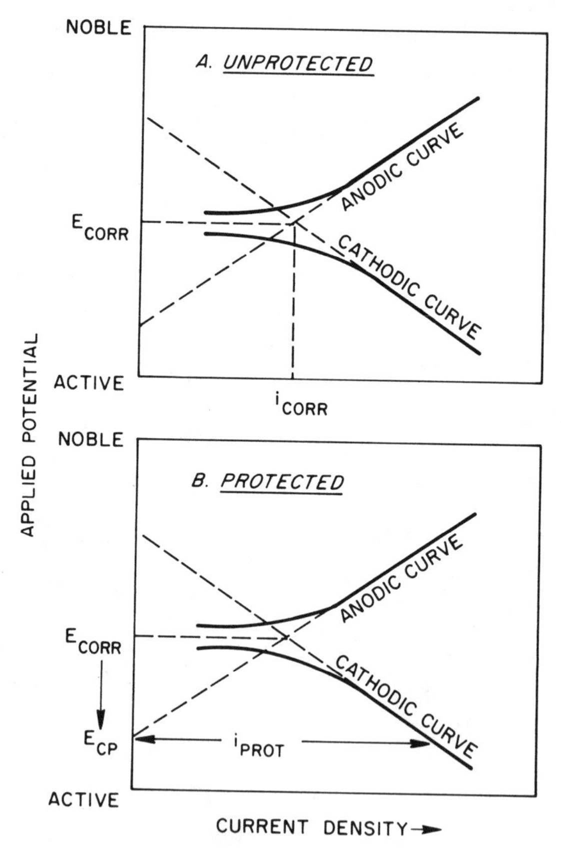

Figure 3-7. Schematic showing principles of cathodic protection.

must be moved from E_{corr} to E_{CP}. This can be achieved by applying a current, i_{prot}, as illustrated in Figure 3-7*b*. The current may be applied from a large DC source or rectified AC source, or by electrically connecting the metal to be protected to a sacrificial anode. Since cathodic protection can be readily used for mild steel, there has been little economic incentive to use it in conjunction with stainless steels in the chemical industry. For the latter, anodic protection is usually preferred, particularly since more aggressive environments can be handled and operating costs for anodic protection are lower than those for cathodic protection (4). However, cathodic protection is used for stainless steels to prevent crevice corrosion and pitting in marine environments, which are discussed in Chapters 5 and 8, respectively.

3-6 APPLICATION OF ELECTROCHEMICAL TECHNIQUES TO EVALUATE LOCALIZED CORROSION

In addition to measuring corrosion rates and defining and developing anodic protection, electrochemical techniques have also found application in the study and evaluation of various forms of localized attack, in both stressed and unstressed stainless steels.

Electrochemical techniques have found extensive application in studies of pitting, because the pitting potential can usually be established by measuring the anodic polarization curve and further refinements achieved by measuring current decay at fixed potentials.

Electrochemical techniques have had somewhat less impact in studies of crevice corrosion, intergranular corrosion, stress corrosion cracking, corrosion fatigue, and hydrogen embrittlement, since unlike in the case of pitting, there are no markedly obvious features of the polarization curve that can be readily associated with these forms of localized corrosion. In the case of crevice corrosion both i_{crit} (Figure 3-5) and certain differences between the polarization curves obtained on forward and reverse scanning have been associated with resistance to crevice corrosion. In the case of intergranular corrosion one of a number of standard techniques employed to evaluate the degree of sensitization, namely, the oxalic acid etch, involves the application of a current. Very recently more complex electrochemical techniques have been proposed to evaluate sensitization.

In stress corrosion studies electrochemical techniques have been used to establish potential regimes in which cracking is likely to occur, to examine the effects of applied anodic and cathodic currents on the initiation and propagation of stress corrosion cracks, and to simulate on a measurable scale the conditions thought to occur at the crack tip in stress corrosion cracking. In studies of corrosion fatigue and hydrogen embrittlement, electrochemical techniques have been used primarily to accelerate or inhibit these forms of attack by the application of anodic or cathodic currents.

The electrochemical studies noted above are discussed in the appropriate chapters dealing with the various forms of localized attack.

REFERENCES

1. C. Wagner and W. Traud, *Z. Electrochem.*, Vol. 44, p. 391, 1938.
2. J. Tafel, *Z. Phys. Chem.*, Vol. 50, p. 641, 1904.
3. P. E. Morris and R. C. Scarberry, *Corrosion*, Vol. 28, p. 444, 1972.
4. M. Henthorne, *Corrosion—Causes and Control*, Carpenter Technology, Reading, Pa., 1973. (Assembled articles reprinted from *Chem. Eng.*, 1971–1972.)
5. S. Evans and E. L. Koehler, *J. Electrochem. Soc.*, Vol. 108, p. 509, 1961.
6. M. Stern and A. L. Geary, *J. Electrochem. Soc.*, Vol. 105, p. 638, 1958.
7. M. Stern and A. L. Geary, *J. Electrochem. Soc.*, Vol. 104, p. 56, 1957.
8. M. Stern and E. D. Weisert, *Proc. ASTM*, Vol. 59, p. 1280, 1959.
9. L. L. Shreir and F. L. LaQue, *Corrosion*, Vol. 2, in *Corrosion Control*, L. L. Shreir, Ed., Newness-Butterworths, Boston, Mass., 1976, p. 20:35.
10. M. G. Fontana and N. D. Greene, *Corrosion Engineering*, McGraw-Hill Book Co., New York, 1967, p. 345.
11. R. L. Leroy, *Corrosion*, Vol. 29, p. 272, 1973.
12. C. Edeleanu, *Nature*, Vol. 173, p. 739, 1954.
13. C. Edeleanu, *Metallurgia*, Vol. 50, p. 113, 1954.
14. R. M. Kain and P. E. Morris, *Anodic Protection of Fe-Cr-Ni-Mo Alloys in Concentrated Sulfuric Acid*, Paper No. 149, presented at NACE Annual Conference, Houston, March 1976.
15. D. A. Jones, reference 9, p. 11.3.

4

PITTING

4-1 INTRODUCTION

Pitting, as the name suggests, is a form of localized corrosive attack that produces pits. It can be a destructive form of corrosion in engineering structures if it causes perforation of equipment. However, minor pitting that does not cause perforation is often tolerated and accepted in engineering equipment for economic reasons. It is not surprising, therefore, that there has been a considerable research effort to develop techniques, mostly ones employing electrochemical methods, capable of defining conditions under which a given metal/environment system is likely to exhibit pitting. For example, Shreir (1) lists some 270 papers published since 1960 dealing with the subject of pitting.

Once pits are initiated they may continue to grow by a self-sustaining mechanism (2). Accordingly, much attention has been paid to obtaining an understanding of the factors controlling their initiation. A number of theories have been proposed that attempt to explain the initiation of pits in perfect surfaces (i.e., surfaces not containing physical defects such as inclusions, compositional heterogeneities, etc.) which consider pit initiation to be a result of certain interactions between discrete species in the environment (e.g., chloride ions) and the passive surface. Among these theories are both kinetic ones, which explain the breakdown of passivity in terms of the competitive adsorption between chloride ions and oxygen, and thermodynamic ones, which consider the pitting potential as that potential at which the chloride ion is in equilibrium with the oxide. For a review and discussion of these and other theories the reader should consult the publications by Shreir (1) and Szklarska-Smialowska (3). The further development of such theories has merit insofar as it is important to obtain a fundamental understanding of the conditions under which pitting can develop in the absence of surface defects. However, commercially

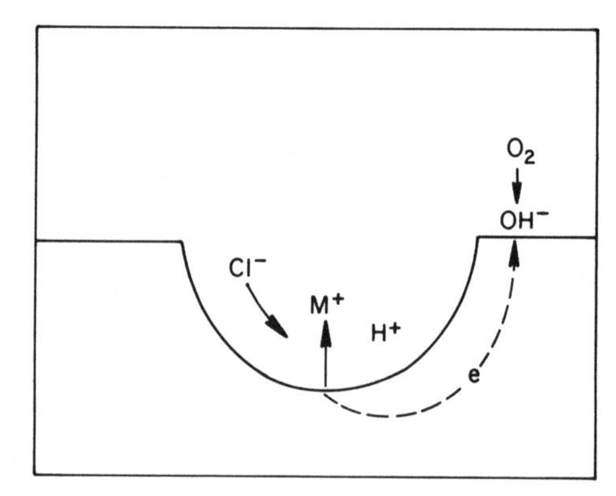

Figure 4-1. A schematic illustration of pit growth mechanism. (After Fontana and Greene.)

produced stainless steels contain numerous inclusions, second phases, and regions of compositional heterogeneity, and there has been a growing body of literature reporting results that identify the initiation of pitting with structural heterogeneities at the surface, particularly, but not exclusively, with inclusions (4–12).

The propagation of pits is thought to involve the dissolution of metal and the maintenance of a high degree of acidity at the bottom of the pit by the hydrolysis of the dissolved metal ions (2). There is a relatively wide acceptance of this propagation mechanism, although all its facets are not fully understood. The pit propagation process is illustrated schematically in Figure 4-1 for a stainless steel pitting in a neutral aerated sodium chloride solution. The anodic metal dissolution reaction at the bottom of the pit, $M \rightarrow M^+ + e$, is balanced by the cathodic reaction on the adjacent surface, $O_2 + 2H_2O + 4e \rightarrow 4OH^-$. The increased concentration of M^+ within the pit results in the migration of chloride ions (Cl^-) to maintain neutrality. The metal chloride formed, M^+Cl^-, is hydrolyzed by water to the hydroxide and free acid,

$$M^+Cl^- + H_2O \rightarrow MOH + H^+Cl^-.$$

The generation of this acid drops the pH values at the bottom of the pit to as low as 1.3 to 1.5 (3) while the pH of the bulk solution remains neutral.

4-2 TECHNIQUES FOR MEASURING TENDENCY FOR PITTING

4-2-1 Electrochemical Techniques

Electrochemical techniques aimed at establishing whether a given alloy/environment combination will give rise to pitting rely on comparing the

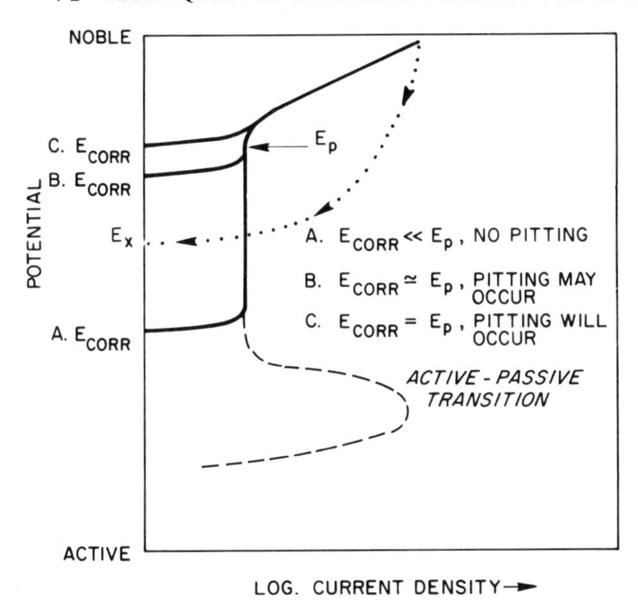

Figure 4-2. Schematic polarization curve illustrating conditions under which pitting may or may not occur.

corrosion potential, E_{corr}, with the pitting potential, E_p, as illustrated in Figure 4-2. Many investigations in the past have measured E_p from the anodic polarization curve (e.g., 13–19).

In acidic aerated aqueous chloride solutions, where the likelihood of pitting of stainless steels is a concern, the corrosion potential, E_{corr}, is often noble to the potential defining the active-passive transition. The anodic polarization curve, therefore, does not pick up the active-passive transition, and the experimental curves are of the type indicated by the unbroken lines in Figure 4-2. Whether pitting will occur depends on the separation of E_{corr} and E_p. If E_{corr} is close to E_p, any small change in the oxidizing power of the solution, such as the introduction of traces of oxidizers (e.g., ferric or cupric ions), can produce pitting by reducing the separation between E_{corr} and E_p. Alloy/environment combinations that are unlikely to cause pitting will have E_{corr} significantly active to E_p, as shown in case A, of Figure 4-2, although obviously noble to the potential defining the active to passive transition in acid environments, otherwise general corrosion will occur. Since the value of E_{corr} of stainless steels in oxygenated chloride solutions may not change significantly from alloy to alloy, it has become customary to equate pitting resistance simply with the absolute value of E_p, rather than $E_p - E_{\text{corr}}$. Thus, it is generally accepted that the more noble the value of E_p, the greater the pitting resistance. However, this could yield misleading results in cases where different alloys exhibit different variations of E_{corr} with time, and the latter should be measured and reported in alloy comparisons.

Sometimes the anodic polarization curve does not resolve as clearly an identifiable current breakout at E_p as the one shown schematically in Figure 4-2. In such cases it can be helpful to refine the estimate of E_p by holding the specimen at various applied potentials E, near E_p, and to monitor the current decay characteristics, as shown in Figure 4-3. When $E = E_p$, the current decays to a value that becomes constant with time.

In the last decade there have been several attempts to introduce the concept of a "protection potential" (against pitting), potential E_x in Figure 4-2, obtained by reversing the direction of potential scanning from some arbitrary current density value. At the present time this concept remains controversial with regard to pitting (20). However, it has been suggested that the parameter $E_p - E_x$, when determined under certain well defined laboratory conditions, can be correlated with susceptibility to crevice corrosion under long time natural marine exposure conditions (20). This observation is discussed further in Chapter 5.

An electrochemical technique, based on identifying E_p as the potential at which a mechanically introduced scratch will repassivate, has been proposed by Pessall and Liu (21). Another technique for measuring pit initiation is that proposed by Morris and Scarberry (22). This technique

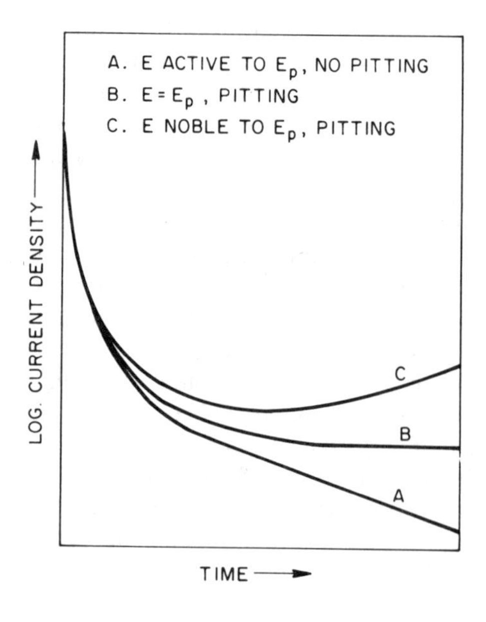

Figure 4-3. Current decay characteristics at various applied potentials (E).

measures E_p by rapid potential scanning and derives its justification both from the empirical observation that for austenitic alloys it gives rise to less noble values of E_p, and that it minimizes the time dependent chances of developing crevice corrosion on the specimen which could cathodically protect the specimen against pitting (23). It should be noted here that many electrochemical techniques for pitting, which employ a slow potential scanning rate, are subject to the conceptual criticism that some crevice corrosion may develop at the specimen/holder or specimen/lacquer interface despite precautions taken to avoid it.

More recently, Lizlovs and Bond (24) have compared pitting potentials of a series of ferritic Fe-Cr-Mo stainless steels in an ambient temperature chloride solution obtained by the following methods: (*a*) the conventional slow potential scanning technique (260 mV/hour); (*b*) the fast potential scanning technique (50 V/hour) proposed by Morris and Scarberry (22); and (*c*) the scratch technique proposed by Pessall and Liu (21). An example of their findings is shown in Table 4-1. Examination of these data, with the view that the least noble pitting potential is likely to be the best indication of susceptibility to pitting, does not identify a preferred technique for ferritic stainless steels. Thus, the slow scanning technique gives the least noble E_p for the molybdenum free Fe-18% Cr alloy, the fast scanning technique gives the least noble E_p for the Fe-18% Cr-2% Mo alloy, and the scratch technique gives the least noble E_p for the Fe-18% Cr-1% Mo and Fe-18% Cr-5% Mo alloys. Since the slow scan technique is time consuming and the scratch technique experimentally more difficult, the fast scan technique would appear to be more desirable from an experimental viewpoint, although the unusually high value for the 5% molybdenum alloy requires further investigation. It should be noted that the comparison was limited to ferritic stainless steels and did not include the more widely used austenitic stainless steels.

It is clear from the foregoing that the determination of pitting potentials

TABLE 4-1. PITTING POTENTIALS* (VOLTS VS. SCE) FOR 18% Cr-Mo FERRITIC STAINLESS STEELS IN 1 *M* NaCl AT 25°C

Stainless Steel	260 mV/h Scan	50 V/h Scan	Scratch Technique
Fe-18%Cr	0.090	0.095	0.120
Fe-18%Cr-1%Mo	0.202	0.195	0.189
Fe-18%Cr-2%Mo	0.297	0.224	0.284
Fe-18%Cr-3.5%Mo	0.423	0.413	0.413
Fe-18%Cr-5%Mo	0.637	0.680	0.535

*Averages of two or more determinations.

by electrochemical techniques is still a subject of ongoing research and discussion. Further comparative studies covering wider alloy ranges and environments and correlations between pitting potential measurements and fixed potential exposures are clearly necessary to generate a greater degree of acceptance of electrochemical techniques. At present, it seems advisable to adopt the electrochemical technique which gives the least noble E_p value for a given alloy/environment system and to spot-check the predictions obtained from anodic polarization curves by long time immersion at fixed applied potentials.

4-2-2 Ferric Chloride Tests

So far the discussion has centered on determining E_p by electrochemical techniques, since the severity (i.e., oxidizing power) of the environment can be varied simply by increasing the potential in the noble direction. A similar increase in severity could be achieved by immersion tests. Conceptually, a series of immersion test solutions could be devised, capable of establishing pitting thresholds, without the use of electrochemical stimulation. However, immersion tests of this type would be open to objection on the grounds that numerous unquantifiable compositional changes would have to be made in the test solutions to vary the oxidizing power. Nevertheless, an immersion test employing ferric chloride (25) has been and continues to be used as a rough screening test for pitting resistance in alloy evaluation. Some variants of this test employ a 10% $FeCl_3 \cdot 6H_2O$ solution acidified with hydrochloric acid, which has an oxidizing power sufficient to pit most stainless steels. Pitting resistance is quantified in terms of readily measurable parameters such as number of pits per unit area and weight loss per unit area. Thus, the ferric chloride test can yield useful comparative information which in many instances correlates with electrochemical parameters. For example, it has been shown (26) that for molybdenum additions to type 301 and sulfur additions to type 316 the ferric chloride test gives a correlation with an electrochemical test, as given in Table 4-2.

A ferric chloride test for pitting, without a hydrochloric acid addition, has been standardized as ASTM Practice G48-76. This test employs a solution containing 100 g of reagent grade ferric chloride, $FeCl_3 \cdot 6H_2O$, in 900 ml of distilled water (27), which yields a solution of about 6% $FeCl_3$ by weight. Exposures are carried out at 22°C or 50°C, the higher temperature providing a more aggressive environment. Evaluation may include photographic reproduction of the pitted specimen surfaces and weight loss measurement. A more detailed examination, outlined in ASTM Practice G46-76, may include measurement of pit density, size, and depth, as shown in Figure 4-4, and characterization of the cross-sectional shape of pits, as illustrated in Figure 4-5 (27).

TABLE 4-2. EFFECTS OF MO ADDITIONS TO TYPE 301 AND S ADDITIONS TO TYPE 316 ON PITTING PARAMETERS DETERMINED BY FERRIC CHLORIDE AND ELECTROCHEMICAL TESTS

Alloy	Ferric Chloride Test*: Pits per in²	Ferric Chloride Test*: Weight Loss, mg/in²	Electrochemical Potential** at 3 mA/cm², Volts vs. SCE
Type 301 +0.03%Mo	14.1	16.3	0.380
+0.26%Mo	3.5	10.6	0.460
+0.51%Mo	2.4	8.3	0.515
+0.75%Mo	0.8	8.3	0.565
Type 316 +0.007%S	0.49	1.7	0.940
+0.017%S	0.22	8.8	0.835
+0.040%S	0.82	5.6	0.820
+0.320%S	1.64	17.3	0.670

*108 g $FeCl_3 \cdot 6H_2O$ + 4.15 cc conc. HCl per litre, 30°C.
**This value does not represent E_p, but is the potential slightly above E_p at a constant current density of 3 mA/cm². Solution was 1N H_2SO_4 + 0.5M NaCl at 30°C.

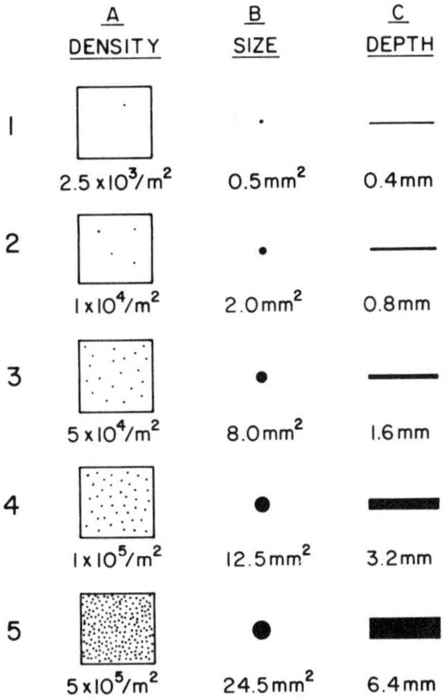

	A DENSITY	B SIZE	C DEPTH
1	2.5 x 10³/m²	0.5 mm²	0.4 mm
2	1 x 10⁴/m²	2.0 mm²	0.8 mm
3	5 x 10⁴/m²	8.0 mm²	1.6 mm
4	1 x 10⁵/m²	12.5 mm²	3.2 mm
5	5 x 10⁵/m²	24.5 mm²	6.4 mm

Figure 4-4. Standard rating chart for pits, ASTM Practice G46-76.

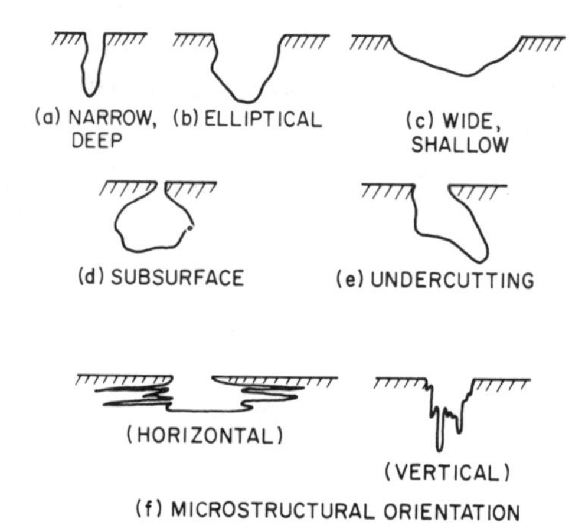

Figure 4-5. Variations in the cross-sectional shape of pits, ASTM Practice G46-76.

4-3 EFFECTS OF COMPOSITION

Horvath and Uhlig (18) have shown that both the major alloying elements of stainless steels, chromium and nickel, increase resistance to pitting, as illustrated in Figures 4-6 and 4-7. However, another important alloying element that increases resistance to pitting is molybdenum (18, 28), as demonstrated in Figure 4-8 (18). While these investigators based their

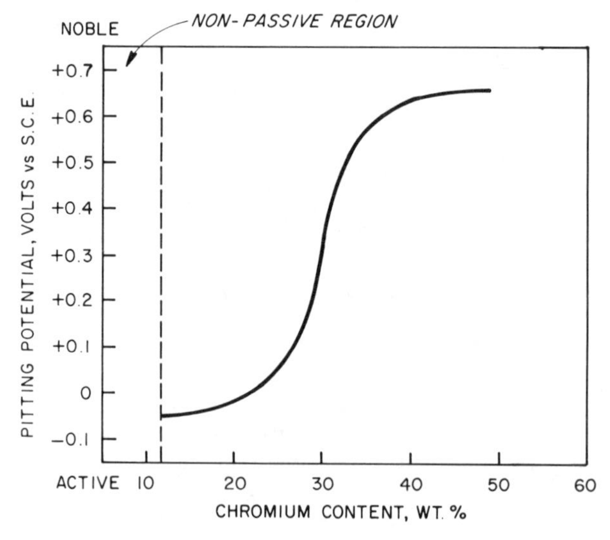

Figure 4-6. Effect of chromium content on pitting potential of iron-chromium alloys in a deaerated 0.1 *N* NaCl solution at 25°C. (After Horvath and Uhlig.)

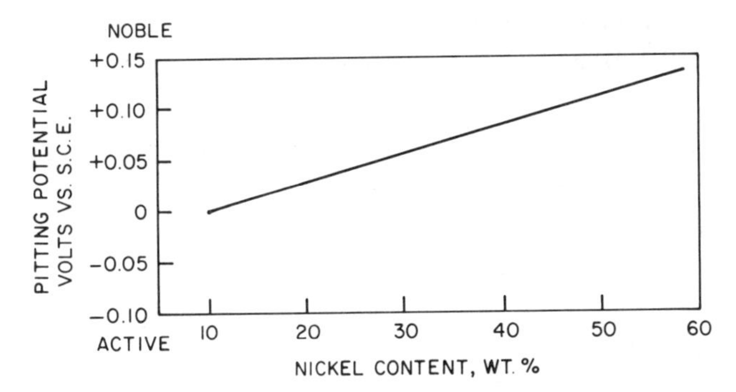

Figure 4-7. Effect of nickel content on pitting potential of Fe-15% Cr alloys in a deaerated 0.1 *N* NaCl solution at 25°C. (After Horvath and Uhlig.)

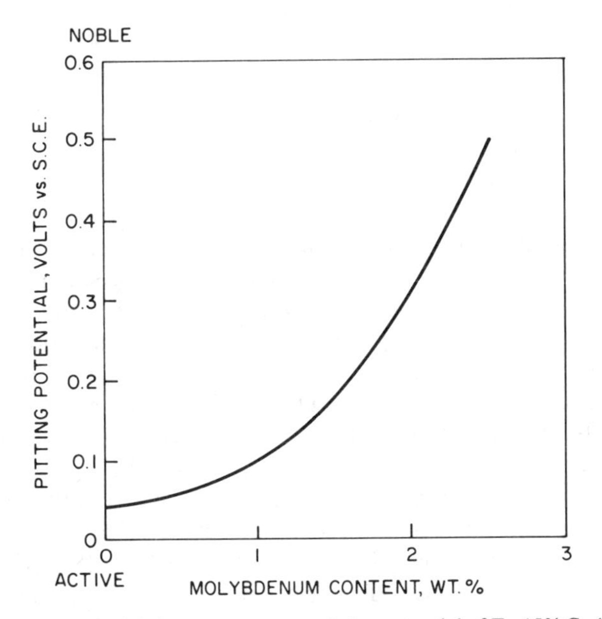

Figure 4-8. Effect of molybdenum content on pitting potential of Fe-15% Cr-13% Ni alloys in a deaerated 0.1 *N* NaCl solution at 25°C. (After Horvath and Uhlig.)

observations on laboratory prepared alloys, the same pattern of increasing pitting resistance with increasing chromium, nickel, and molybdenum contents has been reported by Truman (29) for a series of commercially available British stainless steels, as given in Table 4-3. It should be noted that these three elements also provide the main alloying variants of the highly pitting resistant proprietary alloys listed in Table 2-8.

TABLE 4-3. PITTING POTENTIALS FOR VARIOUS STAINLESS STEELS IN 0.1 *M* NaCl + 0.1 *M* $NaHCO_3$ SOLUTION

Steel*	Category	Nominal Composition % Cr	% Ni	% Mo	Pitting Potential*** (Volts vs. SCE)
410	Martensitic	12.5	--	--	-0.22
430	Ferritic	17.0	--	--	-0.12
431	Martensitic	16.5	2.5	--	-0.12
434	Ferritic	17.0	--	1.1	-0.08
14-5 PH	Prep. Hard.	14.0	5.5	1.6	-0.08
304	Austenitic	18.0	10.0	--	-0.07
315**	Austenitic	17.5	10.5	1.5	0.00
316	Austenitic	17.5	11.5	2.7	+0.14
317	Austenitic	18.5	13.5	3.5	+0.30

*Nearest AISI equivalents to British grades used in test.
**British grade 315S16.
***-ve = active, +ve = noble.

The beneficial effect of molybdenum on pitting resistance has been confirmed by Brigham (30) using an immersion (as opposed to electrochemical) test. In this study a large number of stainless steels and proprietary alloys were exposed to a 10% $FeCl_3$ solution at a given temperature and checked for the initiation of pitting. If none was observed, the temperature was raised by 2.5°C and the materials retested at the higher temperature until pitting was observed. Brigham's test data for stainless steels and some higher alloys are shown in Figure 4-9, with the compositions of most of the materials tested shown in Tables 2-1, 2-8, and 2-9. (Stainless steels designated 18-2 and 16-25-6 in Figure 4-9 contain 19% Cr-2% Mo-0.4% Mn-0.009% C-0.17% Ti and 16% Cr-25% Ni-6% Mo-2% Mn-0.08% C, respectively.) The relationship between pitting resistance and the molybdenum content is clearly in evidence, with the highest critical pitting temperature being exhibited by the highest molybdenum containing alloy tested, namely, AL-6X.

While chromium, nickel, and molybdenum are the main elements required for high pitting resistance, a number of other elements can have a significant effect. These are summarized in Figure 4-10, which is based largely on the compilation by Moskowitz et al. (26). The detrimental effect of the rare earth element gadolinium is noted in a study by Warren (31). In Figure 4-10 elements are described as having a beneficial, detrimental, or

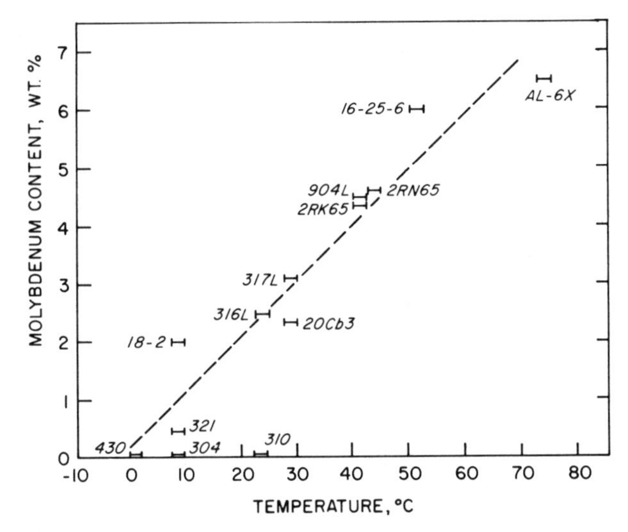

Figure 4-9. Critical pitting temperature in a 10% $FeCl_3$ solution as a function of molybdenum content of a number of commercial stainless steels. (After Brigham.)

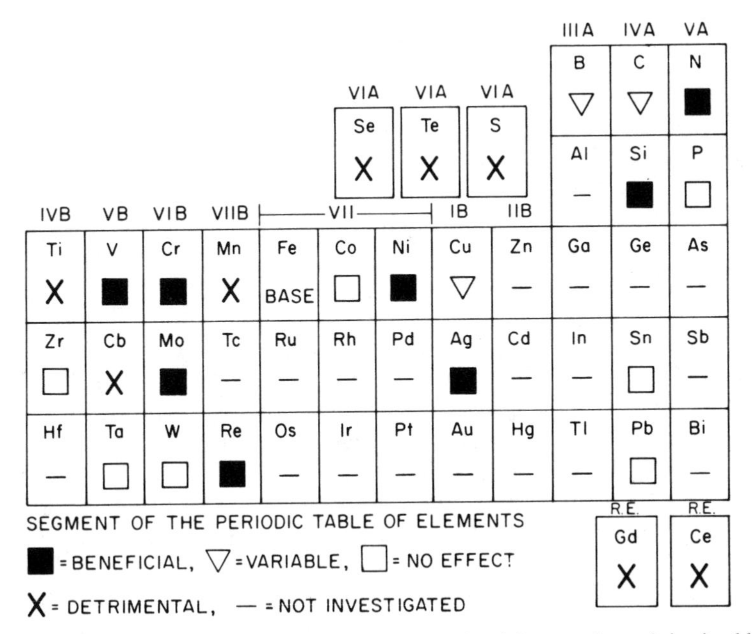

Figure 4-10. Effect of element shown on resistance of stainless steels to pitting in chloride solutions.

variable effect on pitting resistance. Boron, carbon, and copper are described as having a variable effect. Boron is beneficial when in solid solution, but detrimental when precipitated as an intermetallic compound. Carbon has no effect on pitting resistance when in solid solution, but is detrimental when precipitated as a carbide. Copper has no effect in the absence of molybdenum, but a slightly detrimental effect in the presence of molybdenum (26).

4-4 EFFECTS OF MICROSTRUCTURE

As noted in the introductory section of this chapter, microstructure can play an important part in determining pitting resistance. Phases, such as sulfides, delta ferrite, sigma, alpha prime, the strengthening precipitates in precipitation hardening stainless steels, sensitized grain boundaries, and welds, can all have an effect on pitting resistance.

Studies of the role of manganese sulfide inclusions in the pitting of stainless steels have led to a recognition of the importance of the composition of these inclusions. Thus Henthorne (32) has shown that the chromium content of sulfide inclusions increases when the total manganese content of a high sulfur martensitic stainless steel drops below approximately 0.5%, as shown in Figure 4-11. In conjunction with this, other studies (33) have shown that the pitting potentials of various stainless steels in an oxygenated 5% NaCl solution are significantly displaced in the noble direction (indicating increased pitting resistance) when the manganese content is reduced to below approximately 0.4%, as illustrated in Figure 4-12. While extrapolations from high sulfur martensitic to low sulfur austenitic stainless steels may be questionable, there is a strong suggestion here that pitting resistance of stainless steels could be improved by controlling the chromium content of the manganese sulfide inclusions. It has been suggested (34, 35) that at low manganese levels in the steel CrS is the thermodynamically stable sulfide, while above some level of manganese the stable sulfide is an iron-manganese spinel, (FeMn) Cr_2S_4. This spinel appears to be a more effective nucleation site for pitting than CrS. It has also been pointed out (34) that (FeMn) Cr_2S_4 is unstable at temperatures above 1232°C, and hence high temperature annealing could be further investigated as a means of improving pitting resistance as could additions of other sulfide forming elements.

In conjunction with the foregoing, it should be pointed out that the lower pitting resistance of the high manganese type 201 (6.5% Mn) as opposed to that of type 301 (2% Mn maximum), shown in Figure 4-13 (36), is probably less related to sulfide composition and more to the fact that type 201 has a lower nickel content, since Figure 4-12 suggests that there is a little change in pitting potential with manganese contents above approximately 1.6%.

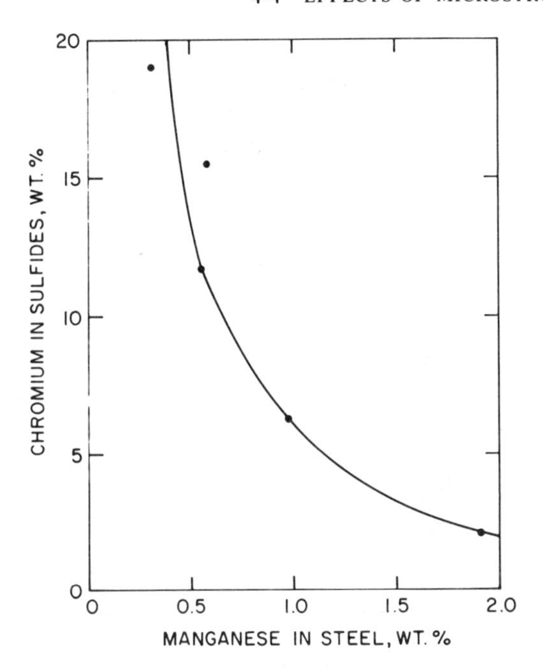

Figure 4-11. Chromium content of in situ sulfides in Fe-13% Cr-0.3% S stainless steels as a function of manganese content. (After Henthorne.)

Delta ferrite in austenitic stainless steels is generally considered detrimental to pitting resistance (5, 37, 39). However, many of the studies aimed at establishing the effect of delta ferrite have been carried out with nitrogen containing steels in which it is difficult to separate the beneficial effect of nitrogen on pitting resistance (26) from its metallurgical tendency to stabilize austenite and thereby prevent the formation of delta ferrite. More recently Dundas and Bond (38) have measured the pitting resistance of an austenitic stainless steel in three conditions of heat treatment, that is, one free of delta ferrite, one containing delta ferrite formed by heat treating at 1345°C, and one in which the delta ferrite formed at 1345°C had been removed by annealing at 1120°C in the austenite region (Figure 2-2). The results, shown in Table 4-4, clearly demonstrate the detrimental effect of delta ferrite on pitting resistance.

Sigma is also detrimental to pitting resistance, at least in ferric chloride environments. Studies by Warren (31) have shown that for type 316L pit initiation times are decreased, and weight loss due to pitting increased as sigma phase is introduced into the microstructure by high temperature heat treatment, as given in Table 4-5. This study did not identify the pitting sites in the microstructure. Sigma contains higher chromium and

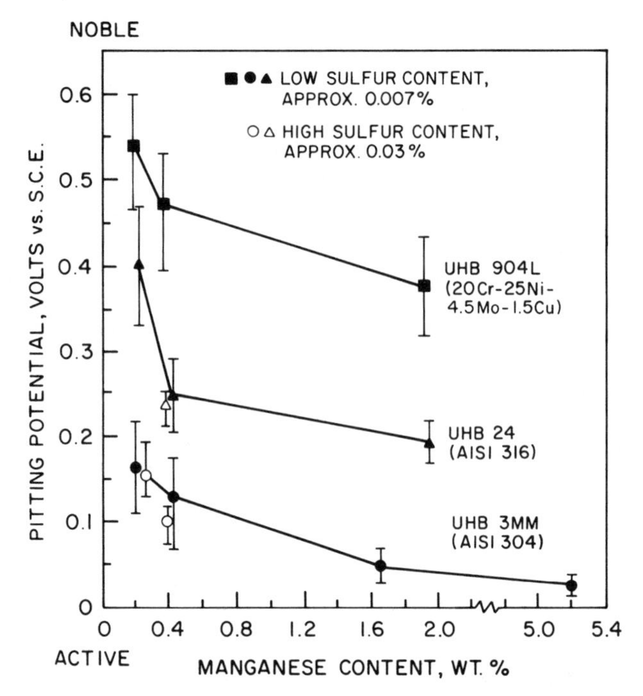

Figure 4-12. Effect of manganese content on the pitting potential of various stainless steels. (After Degerbeck and Wold.)

molybdenum contents than the austenitic matrix and exhibits more noble potentials in chloride solutions than austenite (40). It seems unlikely therefore that the lowered pitting resistance is associated with direct attack on the sigma phase. A more likely explanation may be that some chromium or molybdenum depletion is occurring within the austenite immediately adjacent to the precipitating sigma, and that the pit sites may be associated with such depleted regions. This suggestion is consistent with the observation in Table 4-5 that type 316L heat treated at 870°C for 4 hours exhibited less weight lost by pitting than the material heat treated at 705°C for 1 hour. Rediffusion of chromium or molybdenum back into the depleted regions seems a possible explanation for this observation.

Alpha prime, the phase responsible for the 475°C embrittlement of ferritic stainless steels, also has an adverse effect on pitting resistance. This has been demonstrated (41) for a 19% Cr-2% Mo ferritic stainless steel, as given in Table 4-6.

The effect on pitting resistance of phases precipitated during the strengthening heat treatments given to precipitation hardening stainless

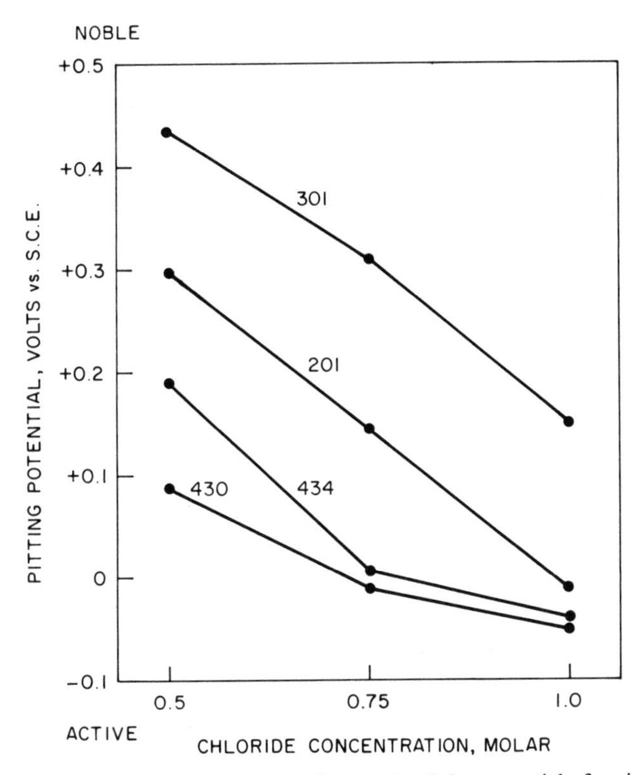

Figure 4-13. Effect of chloride concentration on the pitting potential of various stainless steels in 1 *N* H_2SO_4 solutions. (After Walker and Rowe.)

TABLE 4-4. EFFECT OF DELTA FERRITE PRODUCED BY HEAT TREATING AT 1345°C ON PITTING RESISTANCE OF AN Fe-18% Cr-10% Ni-2.5% Mo-0.16% N ALLOY

Heat Treatment	Volume % Ferrite	Pitting Potential* (Volts vs. SCE)	Corrosion Rate** (mdd)
1120°C/1 h	0	0.30	7.6
1345°C/0.5 h	0.55	0.19	154.0
1345°C/0.5 h + 1120°C/6 h	0	0.29	4.0

*1 M NaCl solution purged with nitrogen, at 45°C.
**10%$FeCl_3 \cdot 6H_2O$ + 0.1N HCl, at 25°C.

TABLE 4-5. EFFECT OF SIGMA PHASE ON PITTING RESISTANCE OF VARIOUS AUSTENITIC STAINLESS STEELS

Material	Heat Treatment	Ferric Chloride Test[a] Time to Initiate Visible Pits (hours)	Weight Loss After 72 hours (g/dm^2)
316L	Mill Annealed	14.5	0.4
316L	705°C/1 h[b]	6.5	4.2
316L	870°C/4 h[c]	5.5	2.2
316 +2.5%Si +0.23%N	870°C/4 h[d]	2.0	12.3
316 +2.5%Si	870°C/4 h[e]	2.0	11.9
317 +2.5%Si	870°C/4 h[f]	2.0	10.5
316 +2.5%Si +0.23%N	Mill Annealed[g]	>72	0

[a] 10% $FeCl_3 \cdot 6H_2O$ at 30°C.
[b] Intergranular network of sigma.
[c] Discrete particles of sigma at grain boundaries.
[d] Discrete particles of sigma at grain boundaries.
[e] 7% original delta ferrite, transformed to sigma.
[f] 25% original delta ferrite, transformed to sigma.
[g] Fully austenitic.

TABLE 4-6. EFFECT OF ALPHA PRIME PHASE ON THE PITTING POTENTIAL OF A 19% Cr-2% Mo FERRITIC STAINLESS STEEL IN 1 *M* NaCl SOLUTION AT 25°C

Heat Treatment	Pitting Potential* (Volts vs. SCE)
815°C/1 h/WQ	+0.338
475°C/ 240 h/WQ	+0.278
475°C/2400 h/WQ	+0.153

*Increasing positive value = increasingly noble.

steels has been evaluated (42) in terms of pitting potential measurements and ferric chloride tests. For the two alloys evaluated, 17-4 PH and Custom 450, the strengthening heat treatment gave rise to slightly more active pitting potentials, indicating slightly lowered resistance to pitting, as shown in Table 4-7. However, these changes appear to be very small both in terms of pitting potential and weight loss in ferric chloride.

TABLE 4-7. PITTING POTENTIALS FOR VARIOUS STAINLESS STEELS IN 3.5% NaCl SOLUTION, pH = 5, AT 24°C, AND WEIGHT LOSS IN 5% $FeCl_3$ AT 24°C AFTER 3 HOUR EXPOSURE

Steel	Category	Heat Treatment	Pitting Potential* (Volts vs. SCE)	Weight Loss, mg
430	Ferritic	Annealed	-0.08	151
410	Martensitic	Annealed	-0.06	--
17-4 PH	Prep. Hard.	Annealed	+0.17	--
		Aged 482°C	+0.16	57
		Aged 621°C	+0.15	62
CUSTOM 450	Prep. Hard.	Annealed	+0.22	35
		Aged 482°C	+0.20	36
		Aged 621°C	+0.16	39
304	Austenitic	Annealed	+0.29	26

*-ve = active, +ve = noble.

The effect of sensitization on pit initiation has been studied by Streicher (5) for austenitic stainless steels and Szklarska-Smialowska and Janik-Czachor (43) for a ferritic stainless steel. Both investigations showed that when present, sensitized grain boundaries act as preferred sites for pit initiation. In martensitic stainless steels the chromium depleted regions around carbides precipitated at certain tempering temperatures are also thought to act as sites for preferential attack (see Chapter 2). As shown in Table 4-8 (44), welds can also provide sensitized areas where deep pitting can occur. In the marine exposure described in Table 4-8, the weld region in type 310 provided the site where pits penetrated through the full

TABLE 4-8. EFFECT OF SEAWATER VELOCITY ON PITTING BEHAVIOR OF WELDED TYPE 316 AND TYPE 310 STAINLESS STEELS (EXPOSURE TIME = 1257 DAYS)

	Velocity = 1.2 m/s			Velocity = 0		
Material	Number of Pits	Max. Pit Depth (mm)	Average Pit Depth (mm)	Number of Pits	Max. Pit Depth (mm)	Average* Pit Depth (mm)
316 Plate	0	0	0	87	1.98	0.96
Weld	0	0	0	47	3.30	1.93
310 Plate	0	0	0	19	2.79	0.96
Weld	0	0	0	23	6.35**	3.05

*Average of 10 deepest pits.
**Perforated through thickness.

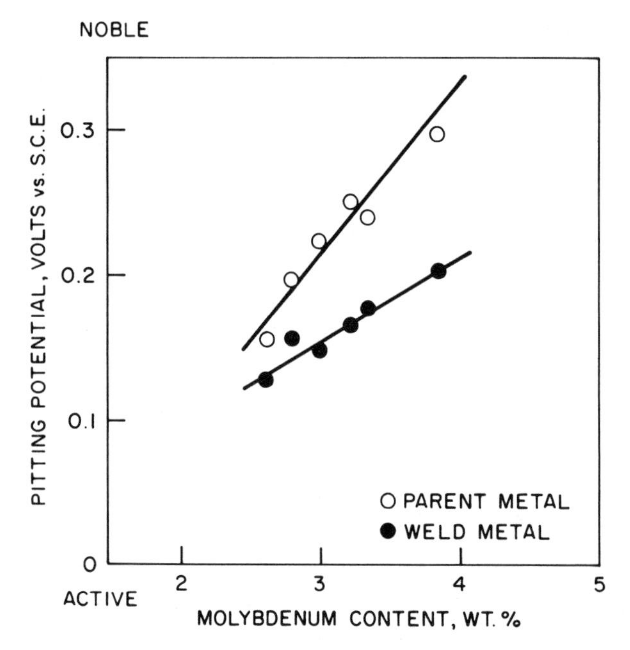

Figure 4-14. Difference between pitting potentials of parent metal and weld metal of a series of Fe-18% Cr-12% Ni-Mo stainless steels in a pH 3, 0.6 *N* NaCl solution at 40°C. (After Garner.)

thickness of the material. In molybdenum containing stainless steels another factor may be important in reducing pitting resistance, namely, molybdenum depletion in the vicinity of delta ferrite formed in welds. Very recent studies (39) of the pitting resistance of a series of welded Fe-18% Cr-12% Ni alloys containing various amounts of molybdenum have shown that the pitting potentials of the weld metal are significantly more active than those of the parent metal, as shown in Figure 4-14, with the difference increasing with increasing molybdenum content. It has been suggested (39) that this effect is associated with preferential pitting attack in molybdenum depleted austenite in the vicinity of delta ferrite precipitated during welding.

In the microstructurally more complex cast stainless steels pitting attack becomes increasingly related to selective attack of the constituent phases. For example, in the ferric chloride tests carried out by Forbes-Jones and Kain (45), summarized in Table 4-9, attack which at low magnifications could be described as pitting was in many cases shown to be a form of selective attack. For the cast alloys CF8M and CD4MCu, the austenite phase was preferentially attacked. In CN7M attack followed

TABLE 4-9. ATTACK OBSERVED IN 10% FERRIC CHLORIDE TESTS OF VARIOUS CAST STAINLESS STEELS AND HIGHER ALLOYS

Cast Alloy	Nearest Wrought Comparative	Maximum Penetration (mm) in 72 h Room Temperature	50°C
CN7M	20Cb-3	2.3	2.3
CF8M	316	1.8	1.8
CD4MCu	329	1.6	1.4
IN-862	AL-6X	0	1.7
CW-12M-2	HASTELLOY alloy C-276	0	Minor

interdendritic boundaries which were lean in chromium and molybdenum, in IN-862 there was some attack in regions adjacent to precipitated intermetallic compounds, and in CW-12M-2 there was some selective etching of an intermetallic compound present in eutectic regions formed during solidification. From a practical viewpoint it should be pointed out that the 10% ferric chloride test is a very severe test used in alloy ranking and that the alloys listed in Table 4-9 do not exhibit pitting in flowing seawater (45).

4-5 EFFECTS OF ENVIRONMENT

From a practical viewpoint most equipment failures due to pitting are caused by chloride and chlorine containing ions, with hypochlorites (found in bleaches) being particularly aggressive (2). Increasing the chloride concentration of a solution significantly increases the tendency toward pitting, as illustrated in Figure 4-15. Of the other halogen ions, bromides will also cause pitting, but fluoride and iodide solutions show little pitting tendencies. Among metal ions cupric, ferric, and mercuric ions in chloride solutions are particularly aggressive.

Among anions that reduce tendency to pitting in chloride solutions, as indicated by a displacement of the pitting potential in the noble direction, are SO_4^{2-}, OH^-, ClO_3^-, CO_3^{2-}, CrO_4^{2-}, and NO_3^- (3). Their inhibiting tendencies depend on their concentrations and the concentrations of chloride in the solution (17).

Measurements of the effect of pH on pitting potential have revealed relatively little effect of pH in the range 1.6 to approximately 10 (3). This may be related to the fact that the hydrolysis reaction in the pit generates its own characteristic acidity, which may be little influenced by the acidity of the bulk solution. In alkaline solutions, however, the pitting potential is significantly displaced in the noble direction, as shown in Figure 4-16 (46), in line with the known inhibiting effect of higher concentrations of OH^- ions (3, 17).

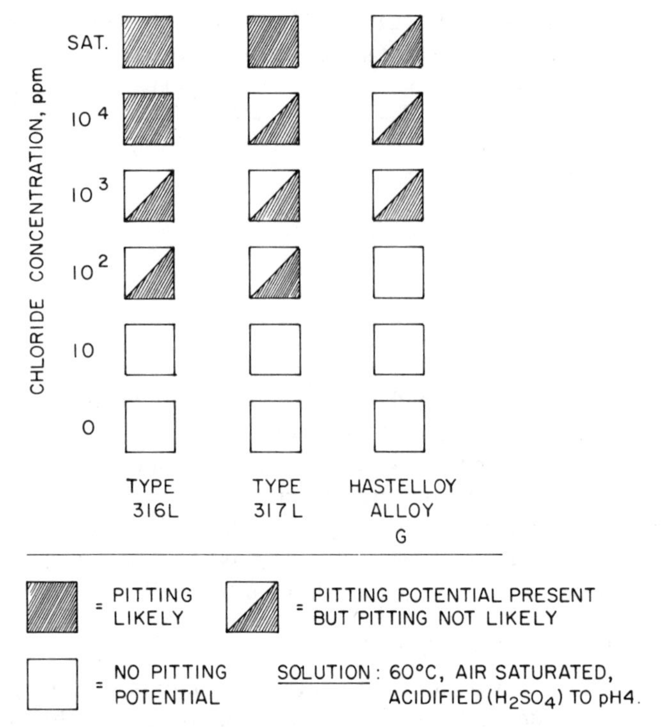

Figure 4-15. Effect of chloride concentration on pit initiation in various austenitic alloys as determined by a comparison of pitting potentials and corrosion potentials.

Increasing the temperature of the solution generally causes the pitting potential to attain more active values which indicates an increased tendency toward pitting, as indicated in Figure 4-17 (46). The effect of temperature on the pitting potential also depends on the composition of the alloy. As seen from Figure 4-17, for the molybdenum containing type 316 at temperatures above 70°C, the pitting potentials become largely independent of temperature.

Since pitting of stainless steels is often associated with stagnant conditions, increasing the velocity of the solution can have a beneficial effect. A practical example (44) of this beneficial effect is illustrated in Table 4-8, where it is seen that seawater flowing at a velocity of 1.2 m/second did not cause attack in welded type 316 and type 310 stainless steel, whereas deep pitting was observed in stagnant seawater. While this attack was identified as pitting, it is possible that it was accelerated by the formation of crevices due to fouling in stagnant seawater.

Another factor which can affect the pitting behavior of stainless steels is dissolved gas in the pitting solution. Although comparatively little work

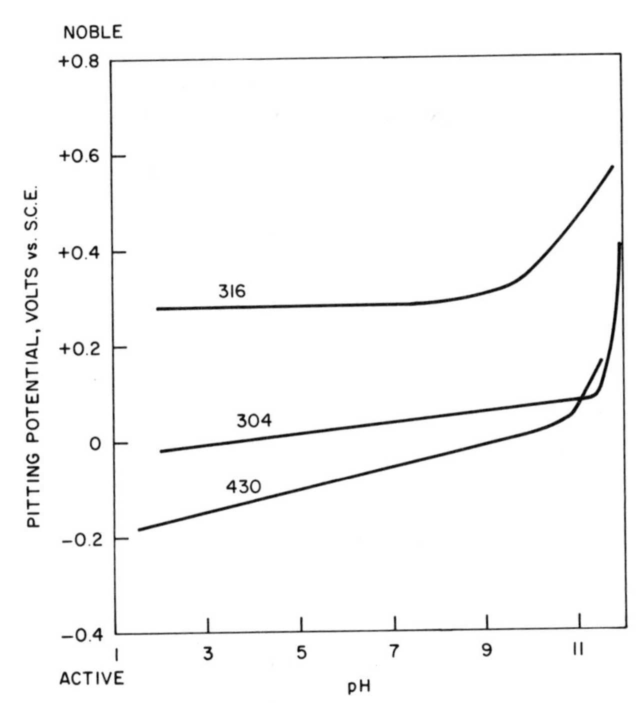

Figure 4-16. Effect of pH on the pitting potential of various stainless steels in a 3% NaCl solution. (After Szklarska-Smialowska.)

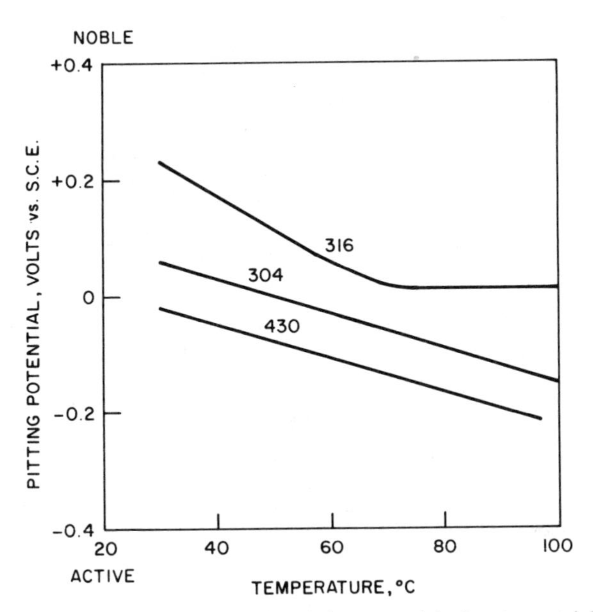

Figure 4-17. Effect of temperature on the pitting potential of various stainless steels in a 3% NaCl solution. (After Szklarska-Smialowska.)

TABLE 4-10. EFFECT OF DISSOLVED GASES ON THE PITTING POTENTIALS OF TYPES 304 AND 430 STAINLESS STEELS IN 1 *M* NaCl SOLUTION AT 25°C

Dissolved Gas	Oxygen Content (ppm)	Pitting Potential* (Volts vs. SCE) Type 304	Type 430
Hydrogen	0.076	-0.050	-0.185
Nitrogen	0.460	-0.020	-0.130
Argon	0.057	+0.050	-0.100
Oxygen	30.1	+0.065	-0.035

*-ve = active, +ve = noble.

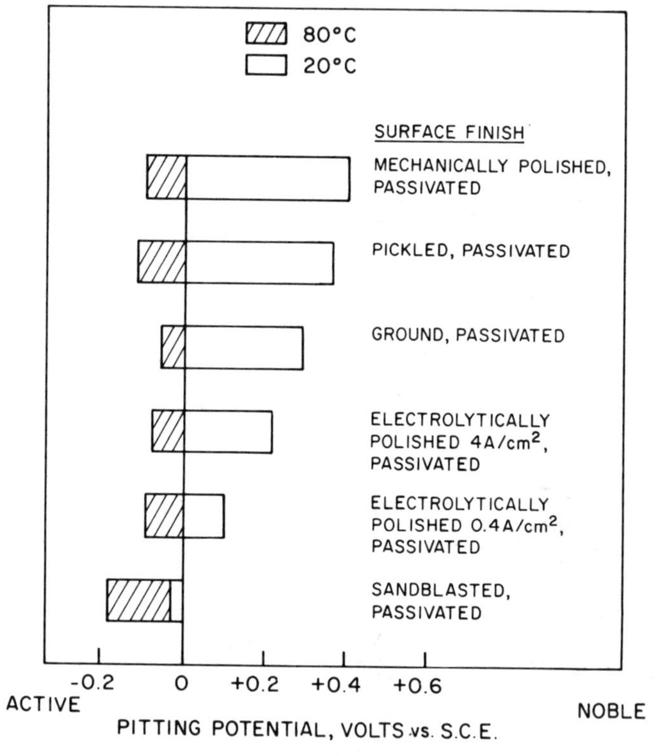

Figure 4-18. Effect of surface finish on pitting potential of type 304 in an aerated 5% NaCl solution. (After Steensland.)

has been done to study this effect, Wilde and Williams (47) have shown that considerable differences in pitting potential can be observed in the presence of different dissolved gases. Their studies, shown in Table 4-10, demonstrate that the most noble potentials are observed when a high concentration of oxygen is present in the solution.

4-6 EFFECTS OF SURFACE CONDITION

A factor in determining pitting potential for a given alloy/environment system is the surface finish of the material. In fact, comparing the pitting tendencies of different alloys having different surface finishes can yield completely meaningless results. A most striking demonstration of the importance of surface finish is that by Steensland (48), illustrated in Figure 4-18, which shows differences in pitting potential in excess of 0.4 V for the same material/environment system using different surface finishes. The "passivation" treatment noted in Figure 4-18 relates to immersion in nitric acid to clean the surface.

Oxide formed by heating stainless steels in air can also have an important effect on pitting resistance. This subject has been studied by Bianchi et al. (49) who examined pitting behavior of type 304L in a glycerol-ethanol solution containing $FeCl_3$, $AlCl_3$, and LiCl, after oxidation in dry air at various temperatures. The results are illustrated in Figure 4-19.

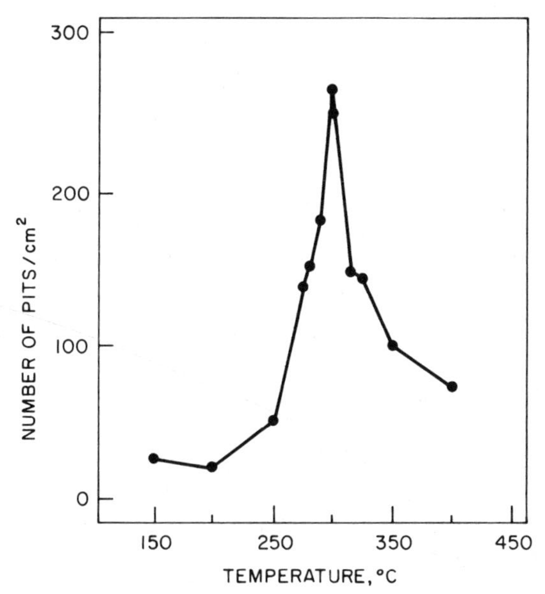

Figure 4-19. The effect of oxidation temperature in dry air on the pit density of type 304L. Oxidized for two hours. (After Bianchi et al.)

The highest pit density was found in type 304L oxidized at 300°C, which has been attributed to changes in the conducting properties of the oxide. Bianchi proposed that high susceptibility to pit nucleation is connected with *n*-type conductivity of the oxide film, whereas low susceptibility to pitting corresponds to *p*-type conductivity. To what extent these observations relate to pitting in aqueous chlorides is not clear. However, there is a certain amount of experience to suggest that oxide films produced by welding can adversely affect pitting resistance in strong chloride solutions. More definitive studies in this area would seem desirable.

Among chemical surface treatments aimed at improving pitting resistance, a well known one is "passivation," which consists of immersing the stainless steel in a 20% HNO_3 solution. The main effect of this treatment is to clean the stainless steel by dissolving surface inclusions and contaminants, such as iron and steel particles embedded in the surface during fabrication. This treatment also removes from the surface manganese sulfide particles which can act as pit initiation sites, and can be used to improve the pitting resistance of the high sulfur free-machining grades of stainless steels (32). To neutralize any residual acid entrapped in the cavities formed by the dissolution of the manganese sulfide particles, the steel after passivation may be rinsed in a sodium hydroxide solution (50).

REFERENCES

1. L. L. Shreir, *Corrosion, Metal/Environment Reactions*, L. L. Shreir, Ed., Vol. 1, Newness-Butterworths, Boston, Mass., 1976, p. 1:182.
2. M. G. Fontana and N. D. Greene, *Corrosion Engineering*, McGraw-Hill Book Co., New York, 1967, p. 51.
3. S. Szklarska-Smialowska, *Localized Corrosion*, National Association of Corrosion Engineers, Houston, Tex., 1974, p. 312.
4. H. H. Uhlig, *Trans. AIMME*, Vol. 140, p. 411, 1940.
5. M. A. Streicher, *J. Electrochem. Soc.*, Vol. 103, p. 375, 1956.
6. K. Lorenz and G. Medwar, *Tyssenforschung*, Vol. 1, p. 97, 1969.
7. B. Forchhammer and H. J. Engell, *Werkst. Korros.*, Vol. 20, p. 1, 1969.
8. B. E. Wilde and J. S. Armijo, *Corrosion*, Vol. 23, p. 208, 1967.
9. S. Steinemann, *Mem. Sci. Rev. Met.*, Vol. 65, p. 615, 1969.
10. M. Smialowski et al., *Corros. Sci.*, Vol. 9, p. 123, 1969.
11. Z. Szklarska-Smialowska et al., *Br. Corros. J.*, Vol. 5, p. 159, 1970.
12. T. P. Hoar et al., *Corros. Sci.*, Vol. 5, p. 279, 1965.
13. N. A. Nielsen and T. N. Rhodin, *Z. Electrochem.*, Vol. 6, No. 2, 1958.
14. J. M. Kolotyrkin, *Corrosion*, Vol. 19, p. 261, 1963.
15. W. Schwenk, *Corros. Sci.* Vol. 5, p. 245, 1965.
16. V. Hospadruk and J. V. Petrocelli, *J. Electrochem. Soc.*, Vol. 113, p. 878, 1966.
17. H. P. Leckie and H. H. Uhlig, *J. Electrochem. Soc.*, Vol. 113, p. 1262, 1966.
18. J. Horvath and H. H. Uhlig, *J. Electrochem. Soc.*, Vol. 115, p. 791, 1968.
19. M. J. Johnson, *Localized Corrosion—Cause of Metal Failure*, ASTM STP-516, American Society for Testing and Materials, Philadelphia, Pa., 1972, p. 262.

20. B. E. Wilde, *Localized Corrosion*, National Association of Corrosion Engineers, Houston, Tex., 1974, p. 342.
21. N. Pessall and C. Liu, *Electrochem. Acta*, Vol. 16, p. 1987, 1971.
22. P. E. Morris and R. C. Scarberry, *Corrosion*, Vol. 28, p. 444, 1972
23. P. E. Morris, *Galvanic and Pitting Corrosion—Field and Laboratory Studies*, ASTM STP-576, American Society for Testing and Materials, Philadelphia, Pa., 1976, p. 261.
24. E. A. Lizlovs and A. P. Bond, *Corrosion*, Vol. 31, p. 219, 1975.
25. H. A. Smith, *Met. Prog.*, Vol. 33, p. 596, 1938.
26. A. Moskowitz et al., *Effects of Residual Elements on Properties of Stainless Steels*, ASTM STP-418, American Society for Testing and Materials, Philadelphia, Pa., 1967, p. 3.
27. *Annual Book of ASTM Standards*, American Society for Testing and Materials, Philadelphia, Pa., 1976.
28. N. D. Tomashov et al., *Corrosion*, Vol. 20, p. 166, 1964.
29. J. E. Truman, *Corrosion, Metal/Environment Reactions,* L. L. Shreir, Ed., Vol. 1, Newness-Butterworths, Boston, Mass., 1976, p. 3:31.
30. R. J. Brigham, *Mater. Performance*, Vol. 13, p. 29, November 1974.
31. D. Warren, "Microstructure and Corrosion Resistance of Austenitic Stainless Steels," Sixth Annual Liberty Bell Corrosion Course, NACE, Philadelphia, Pa., September 1968.
32. M. Henthorne, *Sulfide Inclusions in Steel*, American Society for Metals, Metals Park, Ohio, 1975, p. 445.
33. J. Degerbeck and E. Wold, *Werkst. Korros.* Vol. 25, p. 172, 1974.
34. C. W. Kovach et al., *Trans. ASM*, Vol. 61, p. 575, 1968.
35. M. Henthorne, *Corrosion*, Vol. 26, p. 511, 1970.
36. M. S. Walker and L. C. Rowe, *Corrosion*, Vol. 25, p. 47, 1969.
37. A. P. Bond and E. A. Lizlovs, *J. Electrochem. Soc.*, Vol. 115, p. 1130, 1968.
38. H. J. Dundas and A. P. Bond, *Effects of Delta Ferrite and Nitrogen Contents on the Resistance of Austenitic Steels to Pitting Corrosion*, paper presented at NACE Corrosion/75, Preprint No. 159, 1975.
39. A. Garner, *Proceedings of the International Symposium on Pulp and Paper Corrosion Problems*, Denver, Co., May 1977, to be published by NACE.
40. T. P. Hoar and K. W. J. Bowen, *Trans. ASM*, Vol. 45, p. 443, 1953.
41. E. A. Lizlovs and A. P. Bond, *J. Electrochem. Soc.*, Vol. 122, p. 589, 1975.
42. M. Henthorne et al., *Custom 450—A New High Strength Stainless Steel*, paper presented at NACE Corrosion/72, Preprint No. 53, 1972.
43. Z. Szklarska-Smialowska and M. Janik-Czachor, *Corros. Sci.*, Vol. 7, p. 65, 1967.
44. G. E. Moller, *The Successful Application of Austenitic Stainless Steels in Sea Water*, paper presented at NACE Corrosion/75, Preprint No. 120, 1975.
45. R. M. Forbes-Jones and R. M. Kain, *The Effect of Microstructure on the Corrosion Resistance of Several Cast Alloys*, paper presented at NACE Corrosion/75, Preprint No. 67, 1975.
46. Z. Szklarska-Smialowska, *Corrosion*, Vol. 27, p. 223, 1971.
47. B. E. Wilde and E. Williams, *J. Electrochem. Soc.*, Vol. 116, p. 1539, 1969.
48. O. Steensland, *Anti-Corrosion*, p. 8, May 1968.
49. G. Bianchi et al., *Localized Corrosion*, National Association of Corrosion Engineers, Houston, Tex., 1974, p. 399.
50. M. Henthorne and R. J. Yinger, *Passivation Treatments for Resulfurized Free-Machining Stainless Steels*, ASTM STP-538, American Society for Testing and Materials, Philadelphia, Pa., 1973, p. 90.

5

CREVICE CORROSION

5-1 INTRODUCTION

Crevice corrosion is a form of localized corrosion that can occur within crevices or at shielded surfaces where a stagnant solution is present. Such crevices can be formed at metal/metal or metal/nonmetal junctions, such as those associated with rivets, bolts, gaskets, valve seats, loose surface deposits, or marine growths. Because it can destroy the integrity of mechanical joints in engineering structures, crevice corrosion is generally considered to be a troublesome form of localized corrosion. It is by no means confined to stainless steels and occurs in many alloy systems, including titanium, aluminum, and copper alloys (1), and is of particular concern in marine environments (2). Compared with other forms of localized attack, such as pitting and stress corrosion cracking, crevice corrosion has received relatively little study and is clearly deserving of more attention because of its engineering significance.

To function as a corrosion site, a crevice has to be of sufficient width to permit the entry of the corrodent, but sufficiently narrow to ensure that the corrodent remains stagnant. Accordingly, crevice corrosion usually occurs in gaps a few micrometers wide, and is generally not found in wide grooves or slots in which circulation of the corrodent is possible. However, as yet there are no hard and fast rules as to the dimensions of a crevice which will act as a corrosion site.

Earlier attempts at explaining the mechanism of crevice corrosion involved various types of concentration cells, such as metal-ion, differential aeration, and active-passive; for a discussion and critique of these various mechanisms the reader should consult the publication by France (3). Fontana and Greene (4) have proposed a unified mechanism that incorporates many of the features of the earlier mechanisms and that is applicable to stainless steels.

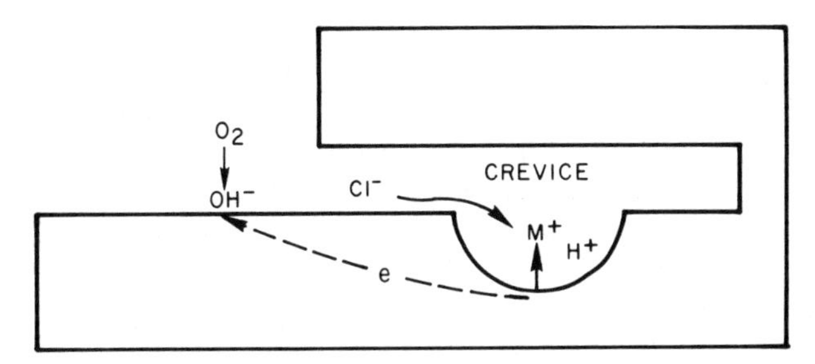

Figure 5-1. A schematic illustration of crevice corrosion mechanism. (After Fontana and Greene.)

This mechnism is very similar in terms of the propagation steps to that proposed for pitting. For a stainless steel undergoing crevice corrosion in an aerated sodium chloride solution, the mechanism suggested by Fontana and Greene is shown in Figure 5-1. This mechanism assumes that initially some corrosion of the stainless steel is occurring in the passive state (Figure 3-3, i_{pass}) uniformly over the entire metal surface, including that surface within the crevice. In terms of the mixed potential theory the anodic reaction, $M \rightarrow M^+ + e$, is balanced by the cathodic reaction $O_2 + 2H_2O + 4e \rightarrow 4OH^-$. However, because the solution within the crevice is stagnant, the oxygen used up in the cathodic reaction is not replenished and the cathodic reaction within the crevice ceases to operate. The anodic reaction, $M \rightarrow M^+ + e$, however, continues within the crevice, building up a high concentration of positively charged metal ions. To balance this charge, negatively charged ions, particularly Cl^-, migrate into the crevice. The resulting metal chloride, M^+Cl^-, is hydrolyzed by water to the hydroxide and free acid,

$$M^+Cl^- + H_2O \rightarrow MOH + H^+Cl^-.$$

This build-up of acidity results in the breakdown of the passive film and a form of corrosive attack similar to autocatalytic pitting. As in the case of pitting, the acid produced by the hydrolysis reaction drops the pH to values below 2 (1), while the pH of the solution outside the crevice remains neutral.

Peterson et al. (5) have attempted to identify the hydrolysis reaction responsible for this degree of acidification by analyzing the species present in natural crevices of type 304 stainless steel and comparing their analytical results with the thermodynamically possible hydrolysis reactions of the alloying constituents of this steel. The results are given in Table 5-1. They concluded that the acidification within the crevice is probably dominated by the hydrolysis of chromic ions, $Cr^{3+} + 3H_2O \rightarrow Cr(OH)_3 + 3H^+$, since only traces of ferric (Fe^{3+}) ions were found analytically, and nickel ions do not hydrolyze to yield pH values significantly below neutral, as shown in Table 5-1. Subsequent studies by Bogar and Fujii (6) have supported the view that the chromic ion determines the acidity attained in stainless steel crevices. This is implied by the fact that pure chromium exhibits the same crevice pH as the stainless steels, as illustrated in Figure 5-2.

TABLE 5-1. EXPRESSIONS FOR LIMITING pH OF HYDROLYSIS REACTION MECHANISMS

Reaction	Equilibrium pH	pH of 1 Normal Solution
$Fe^{2+} + H_2O \rightarrow FeOH^+ + H^+$	pH = 4.75 - 1/2 log $[Fe^{2+}]$	4.75
$Fe^{2+} + 2H_2O \rightarrow Fe(OH)_2 + 2H^+$	pH = 6.64 - 1/2 log $[Fe^{2+}]$	6.64
$Fe^{3+} + 3H_2O \rightarrow Fe(OH)_3 + 3H^+$	pH = 1.61 - 1/3 log $[Fe^{3+}]$	1.61
$Fe^{3+} + 2H_2O \rightarrow Fe(OH)_2^+ + 2H^+$	pH = 2.00 - 1/3 ln $[Fe^{3+}]$	2.00
$2Fe^{3+} + 2H_2O \rightarrow Fe_2(OH)_2^{4+} + 2H^+$	pH = 0.71 - 1/3 ln $[Fe^{3+}]$	0.71
$Fe^{3+} + H_2O \rightarrow FeOH^{2+} + H^+$	pH = 1.52 - 1/2 log $[Fe^{3+}]$	1.52
$Cr^{2+} + 2H_2O \rightarrow Cr(OH)_2 + 2H^+$	pH = 5.50 - 1/2 log $[Cr^{2+}]$	5.50
$Cr^{3+} + 3H_2O \rightarrow Cr(OH)_3 + 3H^+$	pH = 1.60 - 1/3 log $[Cr^{3+}]$	1.60
$Ni^{2+} + 2H_2O \rightarrow Ni(OH)_2 + 2H^+$	pH = 6.09 - 1/2 log $[Ni^{2+}]$	6.09

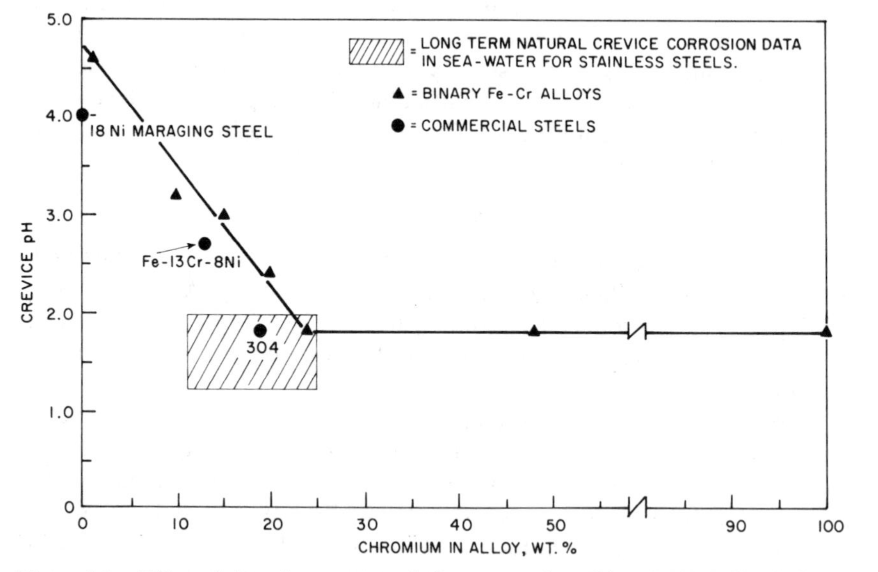

Figure 5-2. Effect of chromium content of alloy on crevice pH in a 3.5% NaCl solution at +0.3 V SCE. (After Bogar and Fujii.)

5-2 TECHNIQUES FOR MEASURING TENDENCY FOR CREVICE CORROSION

5-2-1 Electrochemical Techniques

France (3) has suggested that a distinction be made between "natural" crevice corrosion which initiates in stagnant crevice solutions in the absence of any external electrochemical stimulation as shown in Figure 5-1, and "electrolytic" crevice corrosion which can develop rapidly on electrodes in polarization cells in shielded areas such as electrode-holder joints. Electrolytic crevice corrosion can be detected by its effect on the passive current density, i_{pass}, defined in Figure 3-5. An example of this effect is shown in Figure 5-3 (7), where the electrode mounting techniques designated by the numbers 1, 2, and 3 did not result in a crevice being formed between the electrode and the holder, and hence give the lowest

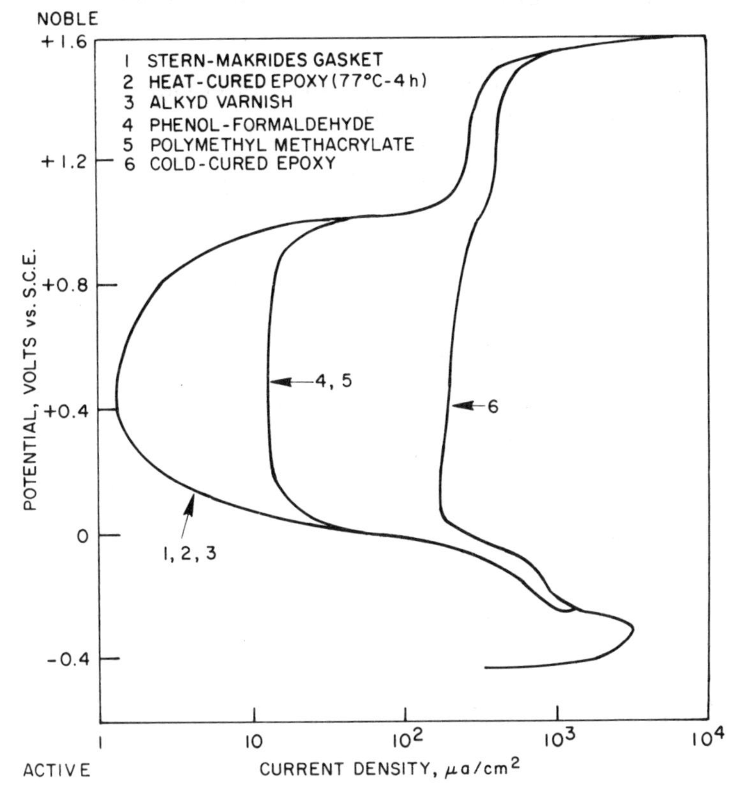

Figure 5-3. Potentiostatic anodic polarization curves of a Fe-10% Cr-10% Ni alloy in a 1 *N* H_2SO_4 solution at 25°C as a function of electrode mounting technique. (After Greene, France, and Wilde.)

values of i_{pass}. However, since the passive current density is also sensitive to a number of other factors such as potential stepping rate and solution composition, the measurement of i_{pass} as a means to evaluate electrolytic crevice corrosion resistance of alloys has not been pursued.

It has been suggested (3) that the electrolytic crevice corrosion resistance of various alloys may be related to the magnitude of their active-passive current densities, i_{crit}, defined in Figure 3-5. The predicted (3) ranking of the electrolytic crevice corrosion resistance in a sulfuric acid solution is given in Table 5-2. It has been pointed out that all the values of i_{crit} listed in Table 5-2 were not determined by the same standard procedure, and that this listing should be regarded as a guideline rather than an inflexible ranking. Nevertheless, it is instructive to compare the ranking derived by comparing i_{crit} in sulfuric acid (Table 5-2), with the ranking determined by Rosenfeld (1) from a study of natural crevice corrosion in a chloride solution. In this study a number of pure metals and U. S. S. R. stainless steels were tested with a crevice formed between the alloy and

TABLE 5-2. PREDICTED ELECTROLYTIC CREVICE CORROSION RESISTANCE IN HYDROGEN SATURATED 1 *N* H_2SO_4 AT 25°C

Material*	i_{CRIT} ($\mu A/cm^2$)
Cr-10%Ni	1
Ti-6%Al-4%V	2
HASTELLOY alloy C	9
Titanium	15
Type 304L	35
Type 347	40
Type 316	45
Type 304	75
Type 201	125
80%Ni-20%Cr	300
Cu-45.4%Ni-2.35%Al	350
CF-8	2,000
Chromium	4,000
Type 430	9,000
Nickel	12,000
Fe-16%Cr	25,000
Iron	120,000
Fe-15%Co	200,000

*Ranked in order of decreasing crevice corrosion resistance.

TABLE 5-3. CREVICE CORROSION OF VARIOUS METALS AND ALLOYS UNDER MEDICINAL RUBBER IN 0.5 *N* NaCl AFTER ABOUT 100 DAYS (RATIO OF OPEN SURFACE TO CREVICE SURFACE = 3:1, SURFACE IN CREVICE = 15 cm^2)

Material	Nominal Composition, Wt. %	Nearest U.S. Equivalent	Area of Corrosion, cm^2	Avg. Rate of Attack, mm/yr
Titanium	--	--	0.00	0.00
Molybdenum	--	--	0.00	0.00
Nickel	--	--	Dark Spots	0.00
Chromium	--	--	0.00	0.00
X18H9M3T*	Fe-18Cr-9Ni-3Mo-Ti	--	0.00	0.00
X18H12M3T	Fe-18Cr-12Ni-3Mo-Ti	--	0.00	0.00
X28	Fe-28Cr	--	0.01	0.22
X18H11B	Fe-18Cr-11Ni-Cb	347	0.13	0.67
X13H4L9	Fe-13Cr-4Ni-9Mn	--	0.90	0.67
X18H15M2B	Fe-18Cr-15Ni-2Mo-Cb	--	0.15	0.74
1X18H9T	Fe-18Cr-9Ni-Ti	321	0.17	0.85
1X18H9	Fe-18Cr-9Ni	--	0.23	1.12
0X18H9	Fe-18Cr-9Ni	304	0.20	1.30
X17	Fe-17Cr	430	0.25	1.52
2X13	Fe-13Cr	410	1.12	1.67

*U.S.S.R. stainless steel designations.

medicinal rubber, the results of which are given in Table 5-3. A comparison of Tables 5-2 and 5-3 reveals some general similarities. For example, in both rankings the ferritic iron-chromium alloys exhibit lower resistance than the nickel and molybdenum containing austenitic stainless steels. However, there is a major discrepancy between the positions of pure chromium and pure nickel in two rankings. In the natural crevice test these metals exhibited virtually no attack, as shown in Table 5-3, whereas the ranking in terms of i_{crit} in sulfuric acid places them low in the order of resistance to crevice corrosion, as given in Table 5-2. Furthermore, Sinigaglia et al. (8) have found that stainless steel rankings in terms of electrolytic crevice corrosion resistance in sulfuric acid solutions depend on the presence of chloride ions. In chloride free sulfuric acid resistance decreased in the order 321 > 316 > 304 > 302, whereas in the presence of chloride resistance decreased in the order 316 > 321 > 302 > 304.

An electrochemical technique for ranking alloys in terms of natural crevice corrosion resistance has been proposed by Wilde (9), who has shown that the potential increment $E_p - E_r$ (defined in Figure 4-2) can be

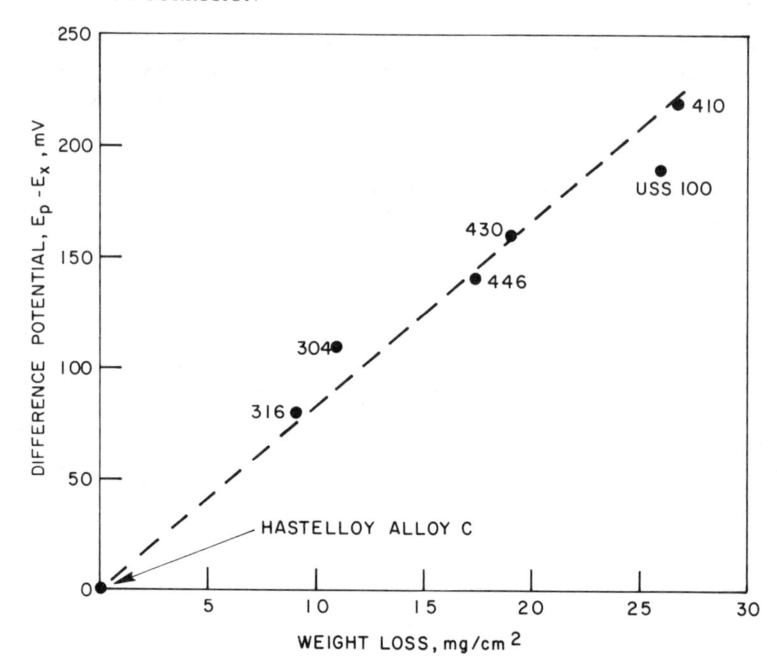

Figure 5-4. Correlation between difference potential and crevice corrosion weight loss of various alloys exposed in seawater for 4.25 years. (After Wilde.)

correlated with natural crevice corrosion weight loss obtained after long term exposures to seawater, as illustrated in Figure 5-4. The parameter $E_p - E_x$ was determined in an aerated 3.5% NaCl solution at 25°C, with E_p being the pitting potential measured at a potential scan rate of 0.6 V/hour and E_x being determined after reversing the scanning direction at a current of 2 mA/cm² (9). The correlation between this laboratory determined parameter and the seawater exposure appears good, as shown in Figure 5-4. Subsequent studies by Rowlands (10) using a similar technique have been less successful in obtaining such a good correlation. In this study crevice corrosion weight loss obtained after a one year exposure in seawater was compared with $E_p - E_x$ measured at a current density of 150 μA/mm², as given in Table 5-4. Both the crevice corrosion weight loss and $E_p - E_x$ values exhibited extremely wide scatter, leading the author to conclude that a statistical approach may be desirable in comparisons of the crevice corrosion resistance of alloys.

It would appear from the foregoing that three features of the anodic polarization curve, namely, i_{pass}, i_{crit}, and $E_p - E_x$, have some relationship to the resistance of stainless steels to natural crevice corrosion in aerated

TABLE 5-4. ASSESSMENT OF CREVICE CORROSION BY EXPOSURE TESTS AND ANODIC HYSTERESIS LOOP METHOD

Nominal Composition, Wt. %	Designation	Condition	Corrosion Rate in Crevice[a], $g/m^2/day$	$E_p - E_x$[b] at $150 \mu A/mm^2$, mV
Fe-18Cr-8Ni-2.5Mo-0.07C	Type 316	Cold Rolled	0.002,0.004,0.046	60,60,60
Fe-18Cr-8Ni-2.5Mo-0.07C	Type 316	Annealed	0.06,0.54,0.54	350,360,240
Fe-17Cr-2Ni-0.16C	Type 431	Tempered 450°C	4.18,3.52,3.52	200,90,130,90
Fe-35Ni-18Cr-9Mo-2Cu	IN-691[c]	Cold Rolled	0.01,0,0	100,110,95
Fe-16Cr-6Ni-1.5Cu-1.5Mo-0.07C	FV520[d]	Soft	1.64,1.78,0.78	270,190
Fe-16Cr-6Ni-1.5Cu-1.5Mo-0.07C	FV520	Peak Hardness	1.12,2.44,1.38	200,160
Fe-16Cr-6Ni-1.5Cu-1.5Mo-0.07C	FV520	Overaged	3.76,1.71,3.00	150,200
Fe-25Cr-5Ni-3Cu-2Mo-0.06C	CD-4MCu (Cast)	Solution Treated and Aged 510°C	<0.01,0.52,0.41	50,120
Fe-25Cr-5Ni-3Cu-2Mo-0.06C	CD-4MCu (Cast)	Solution Treated	0.21,0.10,0.37	120,130

[a] Exposed in seawater at either Langston or Chichester Harbor, England for 1 year.
[b] Oxygen saturated seawater, 20°C, scanning rate 0.15V/min, scan reversed at $600 \mu A/mm^2$.
[c] Designation of The International Nickel Company, Inc.
[d] British precipitation hardening grade.

saline water. Further research aimed at establishing well defined and fundamentally sound correlations between rapid electrochemical laboratory tests and natural crevice corrosion experience covering a broad spectrum of corrosion resistant alloys would seem desirable.

5-2-2 Mathematical Treatments of Crevice Corrosion

As noted in the introductory section of this chapter, the chemical and electrochemical reactions for crevice attack are defined by the Fontana-Greene mechanism. Furthermore, whether a given crevice will act as a localized corrosion site is dependent on its width. However, since gaps between touching materials are determined by asperity contact and the extent of their deformation, it is difficult to measure, reproduce, or even physically visualize the geometry of a tight crevice. No such limitations exist in mathematically idealized crevices. Paradoxically, therefore, despite the unpredictability of its occurrence, crevice corrosion is probably the easiest mathematically quantifiable form of all the types of localized corrosion. Mathematical treatments have been developed for predicting the extent of possible anodic protection against electrolytic crevice corrosion (11) and the location of the maximum attack within a crevice (12), and they are being developed for predicting the time of occurrence (13). Such analyses are particularly valuable since they contribute information about crevice corrosion that is difficult to obtain experimentally.

5-2-3 Natural Crevice Corrosion Tests

To date no standard specimen configuration has been adopted to evaluate the resistance of stainless steels to natural crevice corrosion. Numerous different procedures have been used to form crevices between metal/metal and metal/nonmetal members, varying from sand and debris piled on sheet metal to bolted or riveted joints. For a comprehensive listing and discussion of the various types of crevice specimens that have been used the reader should consult the publications by France (3) and Rosenfeld (1). It should be noted that most test procedures do not attempt to evaluate susceptibility to crevice corrosion on a statistical basis. While it seems unlikely that crevice corrosion is a random form of attack capable of definition only in statistical terms, the fact remains that severe scatter of test results is obtained on nominally identical specimens [e.g., Rowlands (10)]. It is probable that among factors contributing to this scatter may be the inability to obtain reproducible crevice widths, particularly in severe (tight) crevices. As noted before, the width of a tight crevice is determined by asperity contact and the extent of deformation, or in some cases fracture, of the touching asperities. Thus, under certain circumstances it is the surface finish, often not a well defined parameter, which may determine the critical variable of crevice geometry.

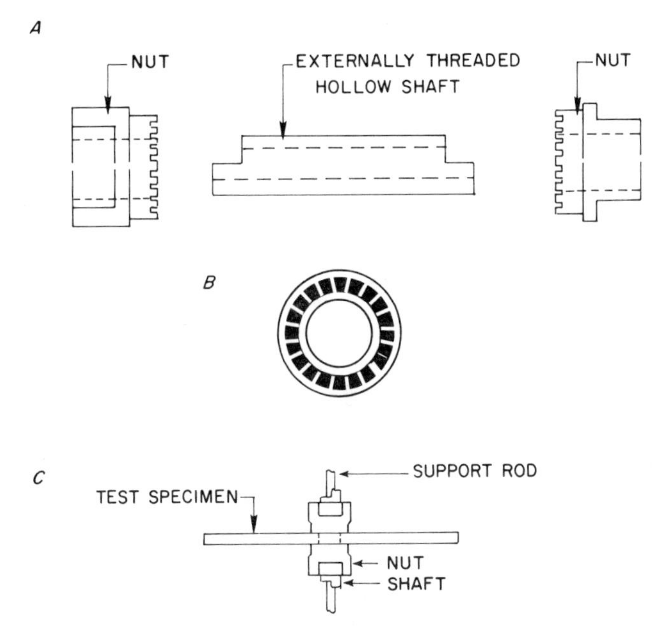

Figure 5-5. Multiple crevice test specimen: *A*, exploded view; *B*, plan view of multiple crevice nut with shaded areas showing crevice sites; *C*, assembled specimen. (After Anderson.)

The crevice corrosion test described by Anderson (14) is a procedure that circumvents this difficulty. In this test a large number of possible crevice corrosion sites are introduced and the merits of the stainless steel under test are assessed on a statistical basis according to the number of crevices which are found to act as localized corrosion sites, and the depth of attack. The crevices are formed between a sheet of stainless steel and grooved Delrin nuts, as exhibited in Figure 5-5. Each nut has 20 plateaus and grooves, providing 40 crevices per specimen. Generally three specimens of each material are tested to provide a statistical base of 120 crevices. After exposure the number of sites attacked and the depth of attack at each site are measured and plotted on arithmetic probability paper (ASTM G16-71). Such plots provide information regarding the probability of crevice corrosion initiation and the probability of corrosive attack to any given depth. A typical example comparing two cast stainless steels CA-15 (410) and CF-8M (316) in seawater is shown in Figure 5-6. These data show that for CA-15 there is a 100% probability of crevice corrosion initiation and a 50% probability that crevice attack will penetrate to depths greater than 0.75 mm. For CF-8M there is a 25% probabil-

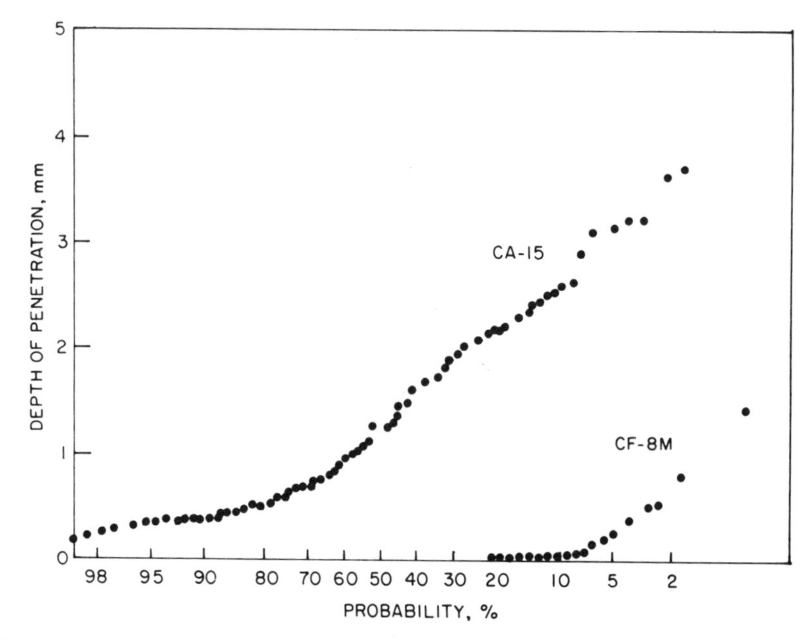

Figure 5-6. Probability of crevice corrosion for cast stainless steels. Exposure conditions: seawater, 32 days, 15°C, 75:1 bold/crevice area ratio. (After Anderson.)

ity that crevice corrosion will initiate, but only a 2% probability that crevice attack will penetrate to depths greater than 0.75 mm.

Studies by Rowlands (10) have suggested that the multiple crevice test does not yield severe crevices, since more attack was found within the crevices formed between the suspension sleeves and washers used to attach the specimens to the exposure panel than under the multiple crevice nuts. However, this problem can be overcome by using the support rod for attachment, as illustrated in Figure 5-5. More recently it has been discovered that the torque used in tightening the multiple crevice nuts can influence results and that consistently reproducible results can be obtained by applying a torque of 10.9 J (8 ft-lbs) in tightening. While such further refinements of test procedure are necessary, it seems that the multiple crevice approach provides a good basis for the development of a natural crevice corrosion test capable of statistical quantification.

5-2-4 Ferric Chloride Test

Another test in which several crevice sites on the same specimen are evaluated is the ferric chloride crevice corrosion test, which has been standardized as ASTM Practice G48-76 (Method B) (15). In this method

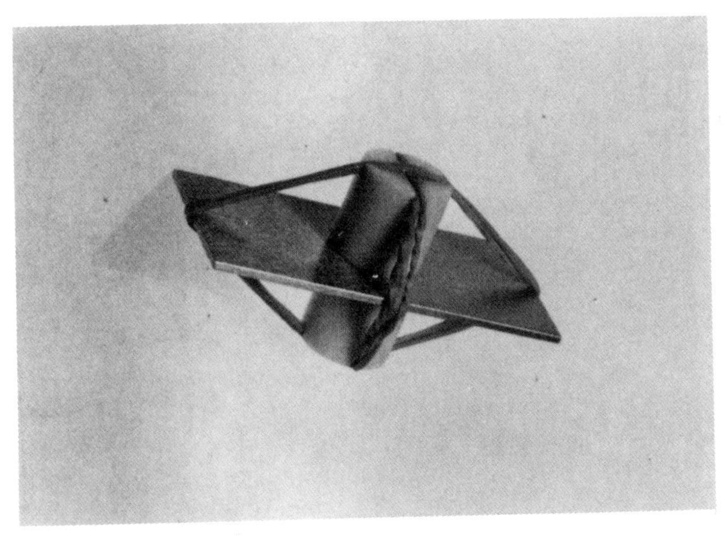

Figure 5-7. Crevice corrosion test specimen used in ASTM Practice G46-76, Method B.

six crevice sites are evaluated, two formed between TFE-fluorocarbon blocks and a sheet specimen and four formed between rubber bands and a sheet specimen edges, as shown in Figure 5-7. The test specimens are immersed in a solution prepared by dissolving 100 g of reagent grade ferric chloride, $FeCl_3 \cdot 6H_2O$, in 900 ml of distilled water. Exposures are carried out at 22°C or 50°C for 72 hours. The average depth of attack under the TFE-fluorocarbon blocks and under the rubber bands is measured and recorded. Recent studies by Kain* suggest that results obtained using this test do not correlate with observations made in heated seawater.

5-3 EFFECTS OF COMPOSITION

There have been very few studies of crevice corrosion resistance that have examined the effects of alloying elements in stainless steels as single variables. Therefore, it has been the practice to infer the effects of various alloying elements by comparing commercial alloys. However, until such time that better data become available, there is little choice but to rely on the guidance obtained this way. With this reservation in mind, crevice corrosion tests on commercial alloys indicate that, as in the case of pitting resistance, the major alloying elements, nickel and chromium, increase

*R. M. Kain, *Crevice Corrosion Resistance of Austenitic Stainless Steels in Ambient and Elevated Temperature Seawater,* Paper No. 230, presented at NACE Annual Conference, Atlanta, March 1979.

TABLE 5-5. CREVICE CORROSION WEIGHT LOSS OF VARIOUS STAINLESS STEELS IN AMBIENT TEMPERATURE SEAWATER EXPOSED FOR 109 DAYS. BOLD AREA 1866 cm^2, SHIELDED AREA 30.5 cm^2, FOUR CREVICES PER SPECIMEN

Steel	Crevice Corrosion Weight Loss, grams
Type 430	42.0
Type 302	16.5

resistance to crevice corrosion. Another important alloying element which increases resistance to crevice corrosion is molybdenum.

The beneficial effect of nickel can be inferred from a comparison of the crevice corrosion resistance of types 430 and 302 given in Table 5-5 (16). These alloys have comparable chromium and other alloying element contents and differ primarily in nickel content. Unless it can be shown that the ferritic structure itself (type 430) is inferior to the austenitic structure (type 302) in terms of crevice corrosion resistance, the data of Table 5-5 must be interpreted as showing a significant beneficial effect of a nickel addition. Such a beneficial effect would be expected from the observation that increasing the nickel content raises the pitting potential in the noble direction, as exhibited in Figure 4-7, thereby rendering the breakdown of the passive film within the crevice more difficult. Also, as shown in Table 5-1, the hydrolysis of Ni^{2+} yields essentially a neutral pH, so that the dissolved nickel ions are not likely to contribute to the acidification of the solution within a crevice.

The beneficial effect of increasing chromium content can be inferred from the data shown in Table 5-6 (17), where it is seen that type 446 (26% Cr) is more resistant to crevice corrosion in seawater than type 430 (17.5% Cr). This effect is also evident in the data shown in Table 5-7 (18) for crevice corrosion of high purity ferritic stainless steels in acidified ferric chloride solutions at various temperatures. It can be seen that at a constant molybdenum level of 2%, increasing the chromium content from 20% to 28% results in a progressive increase in the temperature at which crevice attack is observed. As in the case of nickel, the beneficial effect of increasing the chromium content is probably attributable to the fact that chromium raises the pitting potential in the noble direction (Figure 4-6), rendering film breakdown within the crevice more difficult. However, unlike the nickel ion, the chromic ion hydrolyses to yield a pH of about 1.6, as given in Table 5-1, and, as noted previously, is thought to be the species responsible for the acidification that occurs in stainless steel crevices. It should be noted here that two of the most crevice corrosion

TABLE 5-6. PROBABILITY OF CREVICE CORROSION INITIATION AND MAXIMUM DEPTH OF ATTACK OF VARIOUS STAINLESS STEELS EXPOSED FOR 30 DAYS TO SEAWATER AT 15°C. MULTIPLE CREVICE TECHNIQUE, 120 CREVICES PER MATERIAL

Steel	Composition, Wt. %			Probability of Crevice Corrosion Initiation, %	Maximum Depth of Attack, mm
	Cr	Ni	Mo		
Type 430	17.5	0.26	0.02	52	1.17
Type 446	25.9	0.03	--	7	0.23
Type 304	18.4	9.4	--	13	0.28
Type 316	17.5	13.5	2.3	2	0.03
Type 317	19.0	13.5	3.8	0	0

TABLE 5-7. CREVICE CORROSION TESTS IN 10% $FeCl_3 \cdot 6H_2O$ + 0.1 *N* HCl. WEIGHT LOSS (mg/dm^2) PER DAY COMPUTED OVER ENTIRE COUPON AREA (TEST DURATION IN DAYS GIVEN IN PARENTHESES)

Nominal Composition, Wt. %	Test Temperature, °C			
	25	30	40	50
Fe-20Cr-2Mo	64(1)	197(1)	--	--
Fe-22.5Cr-2Mo	0(7)	3(1)	--	--
Fe-25Cr	670(2.5)	--	--	--
Fe-26Cr-1Mo	0(23)	6.2(4)	229(1)	764(1)
Fe-25Cr-2Mo	0(7)	0(1)	0(1)	210(1)
Fe-28Cr-2Mo	0(7)	0(4)	0(1)	13(1)
Fe-25Cr-3.5Mo	0(24)	--	--	7(5)
Fe-25Cr-5Mo	0(63)	--	--	0(11)

resistant higher alloys, Inconel alloy 625 and Hastelloy alloy C-276, contain only 22% and 16% chromium, respectively, and rely to some degree on their nickel and molybdenum contents to provide their high crevice corrosion resistance.

Among all the alloying elements found in stainless steel, it is undoubtedly molybdenum that provides the greatest improvement in crevice corrosion resistance. There are numerous publications that attest to this fact, irrespective of whether testing employed electrochemical (1, 9, 10, 18), ferric chloride (19–21) or natural seawater exposures (2, 10, 17), and irrespective of whether the alloys are austenitic or ferritic. For austenitic stainless steels a significant improvement in crevice corrosion resistance

TABLE 5-8. CREVICE CORROSION WEIGHT LOSS OF VARIOUS STAINLESS STEELS AND HIGHER ALLOYS TESTED FOR 72 HOURS IN ROOM TEMPERATURE 10% $FeCl_3$ SOLUTION. CREVICES FORMED BY TIGHT RUBBER BANDS

Material	Mo Content, Wt. %	Weight Loss, grams
Type 304	--	0.4 -0.8
Type 316	2.5	0.2 -0.4
Type 317	3.5	0.15-0.2
AL-6X	6.5	0
INCONEL alloy 625	9.0	0
HASTELLOY alloy C-276	16.0	0

in seawater can be obtained by molybdenum additions as low as 2 to 4%, as exemplified by types 316 and 317, as given in Table 5-6. Higher molybdenum proprietary alloys are generally resistant to ambient temperature seawater, as illustrated in Figure 5-4, or ambient temperature ferric chloride, as given in Table 5-8 (19), and have been ranked in heated ferric chloride, as shown in Figure 5-8 (20).

Molybdenum, like nickel and chromium, also raises the pitting potential of the alloy in the noble direction (Figure 4-8), so a likely explanation for the beneficial effect of molybdenum can be proposed in terms of increased difficulty in breaking down the passive film. However, it is possible that

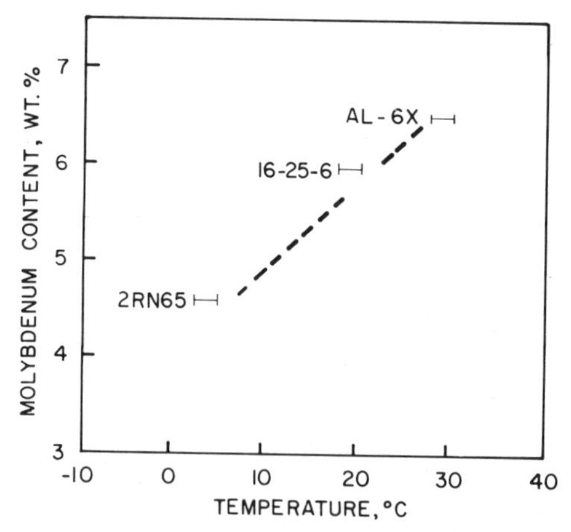

Figure 5-8. Critical crevice corrosion temperature in a 10% $FeCl_3$ solution as a function of molybdenum content for a number of alloys. (After Brigham.)

there is also another effect at higher molybdenum contents that inhibits the propagation, rather than the initiation of crevice attack, namely, the formation of a molybdate salt film. Recent studies by Ambrose (22) of the repassivation kinetics of Fe-Mo alloys have suggested that when the molybdenum content in the alloy exceeds 5%, a protective chloride salt film stabilized by molybdate may precipitate from the solution after some molybdenum has dissolved. It remains to be established whether such precipitation of a protective molybdate salt can occur in stainless steel crevices, pits, or stress corrosion cracks. For chloride stress corrosion this concept could rationalize the observation that stress corrosion cracking resistance increases with molybdenum content only when certain levels of molybdenum are present in stainless steels (23).

Regarding elements other than nickel, chromium, and molybdenum, Degerbeck (24) has shown that silicon, nitrogen and copper, when present in molybdenum containing austenitic stainless steels, have a beneficial effect on crevice corrosion resistance in seawater, as given in Table 5-9. It is not known whether these elements have a beneficial effect in stainless steels without molybdenum. No effect of columbium and titanium on crevice corrosion resistance was found in these seawater tests (24). Streicher (25) has shown that small additions of palladium and rhodium to Fe-28.5% Cr-4.0% Mo alloys are detrimental to crevice corrosion resistance in 50°C ferric chloride solution.

The effects of the various alloying elements on the crevice corrosion resistance of stainless steels are summarized in Figure 5-9. It should be noted that the major alloying elements of stainless steels, nickel, chromium, and molybdenum, have a beneficial effect on both pitting resistance (see Figure 4-10), and crevice corrosion resistance, as illustrated in Figure 5-9. There are insufficient data for crevice corrosion resistance to extend such a comparison to the effects of other alloying elements.

5-4 EFFECTS OF MICROSTRUCTURE

There have been very few evaluations of crevice corrosion resistance of stainless steels that have examined the effects of alloy structure. Studies by Degerbeck (24) of duplex stainless steels have identified austenite/ferrite boundaries as susceptible sites for the initiation and propagation of crevice attack in seawater exposures. As shown in Table 5-9, the duplex stainless steels exhibited deep crevice attack. Studies by Rowlands (10) suggest that precipitation hardening may have a slightly detrimental effect on crevice corrosion resistance, as exhibited in Table 5-4, although the wide scatter of data prevents any firm conclusions to be drawn.

TABLE 5-9. CREVICE CORROSION OF VARIOUS STAINLESS STEELS EXPOSED TO SEAWATER FOR 1 YEAR*

Supplier Designation	Nominal Composition, Wt. %	Nearest U.S. Equivalent	Maximum Depth of Attack, mm
UHB3MM	Fe-18Cr-9Ni	304	>1.5
UHB24L	Fe-18Cr-12Ni-3Mo	316L	1.0
UHB24LN	Fe-18Cr-13Ni-3Mo-N	--	0.3
UHB24LN+Cb	Fe-18Cr-13Ni-3Mo-N-Cb	--	<0.1
UHB624	Fe-18Cr-13Ni-3Mo-Cb	--	1.0
UHB34	Fe-17Cr-14Ni-4Mo	317	1.0
UHB34L	Fe-17Cr-14Ni-4Mo	317L	0.5
--	Fe-18Cr-18Ni-2Si	--	0.8
--	Fe-18Cr-14Ni-3.5Si	--	2.0
--	Fe-17Cr-15Ni-2Mo-3.5Si	--	0.2
--	Fe-18Cr-12Ni-2Si-N	--	1.4
--	Fe-18Cr-14Ni-2Mo-2Si-N	--	<0.1
--	Fe-18Cr-24Ni-4Mo	--	0.8
--	Fe-18Cr-20Ni-4Mo-N	--	0.4
UHB904L	Fe-20Cr-25Ni-5Mo-1.5Cu	--	<0.1
--	Fe-18Cr-24Ni-3Mo-2Cu-Ti	--	<0.1
UHB44	Fe-25Cr-5Ni-1.5Mo (Duplex)	--	2.5
--	Fe-19Cr-5Ni-3Mo-1Si (Duplex)	--	1.7
--	Fe-25Cr-5Ni-1.5Mo-N (Duplex)	--	2.0
--	Fe-25Cr-5Ni-1.5Mo-N (Duplex)	--	1.2

*Specimens exposed in seawater on the west coast of Sweden.

If there is a correlation between pitting and crevice corrosion, as suggested by the similarity of the effects of the major alloying elements on both types of attack (cf. Figures 4-10 and 5-9), then the presence of inclusions and second phases, such as delta ferrite, alpha prime, sigma, sulfides, and hardening precipitates, may have some effect on crevice corrosion behavior. Further evaluations of the effects of microstructure on crevice corrosion resistance would seem desirable.

5-5 OTHER FACTORS

Other factors that have received some attention in studies of crevice corrosion of stainless steels include cathode-to-anode area ratio, cathodic protection, and temperature and velocity of the corrodent.

It is well known (26) that for stainless steel crevices in seawater, decreasing the crevice area or increasing the area of the material outside the crevice results in an increase in crevice attack. The area within the

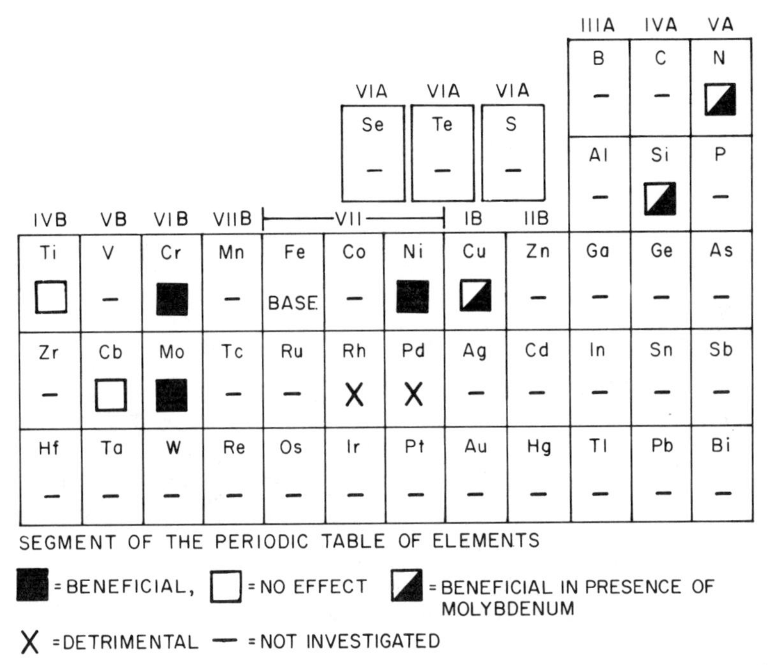

Figure 5-9. Effect of element shown on resistance of stainless steels to crevice corrosion in chloride solutions.

crevice is considered to be anodic and the area outside the crevice to be cathodic, and the above noted area effect has been explained by drawing analogies with the behavior of dissimilar metal galvanic cells (2). In such cells an unfavorable area ratio consists of a large cathode and a small anode. For a given current flow in such a cell, the current density is greater on a small electrode than on a larger one. Consequently, a small anode will have a greater current density and hence a greater corrosion rate than a large anode. The importance of the cathode (bold)-to-anode (crevice) area ratio on the probability of crevice corrosion initiation for type 304, type 316, and Incoloy alloy 825 in seawater is shown in Figure 5-10 (14).

The fact that crevice corrosion of stainless steels in seawater can be suppressed by cathodic protection has also been known for a considerable time (27). However, this technique appears practical only for austenitic stainless steels, since hydrogen blistering results if cathodic protection is attempted for martensitic (type 410) and ferritic (type 430) stainless steels (27). More recent studies have been aimed at establishing the depths of crevices that can be cathodically protected. For example, Lennox and

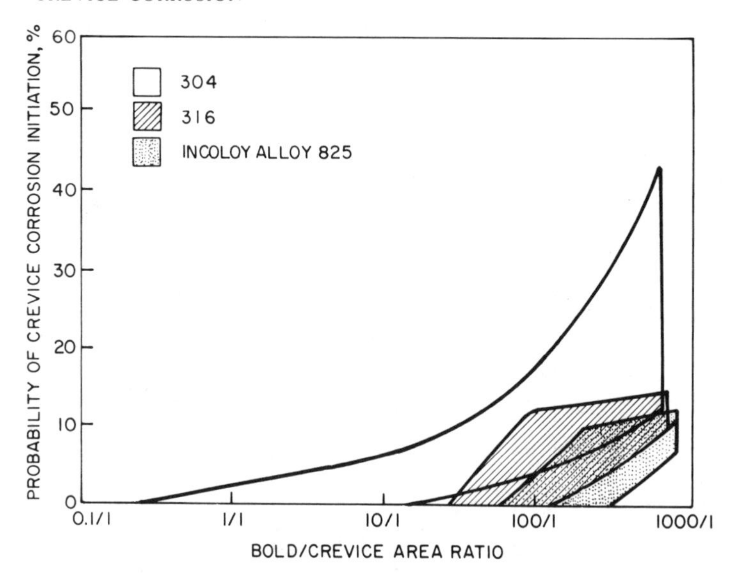

Figure 5-10. Probability of crevice corrosion initiation as a function of bold/crevice area ratio. Exposure conditions: 30 days in flowing ambient temperature seawater. (After Anderson.)

Peterson (28) have shown that major changes of pH to alkaline values, which are indicative of cathodic behavior or the diffusion of alkaline solution generated at cathode areas, can be detected in 0.06 mm crevices to depths exceeding 89 cm (about 3 ft.), when a type 304/acrylic plastic crevice is coupled to a zinc anode, as illustrated in Figure 5-11. Subsequent studies (29) have evaluated the effectiveness of cathodic protection of type 304 tubes and pipes rather than crevices. However, it has been shown that crevice corrosion in seawater at crevices formed by type 304 O-ring seals can be reduced by cathodic protection (5).

The effect of increasing temperature on crevice corrosion is not easy to predict, since it could affect various relevant factors in different and opposing ways. While transport processes and reaction kinetics would be accelerated by increased temperature, the solubility of oxygen would be decreased, and in neutral aerated chloride solutions it is the reduction of oxygen, $O_2 + 2H_2O + 4e \rightarrow 4OH^-$, that provides the main cathodic reaction on the surface outside the crevice. Among other factors that could be affected are the pitting potential, hydrolysis equilibria, and the composition, structure, and properties of surface films. With regard to the latter, Efird (17) has shown that the magnesium content of surface films is increased by increasing temperature for stainless steels exposed to

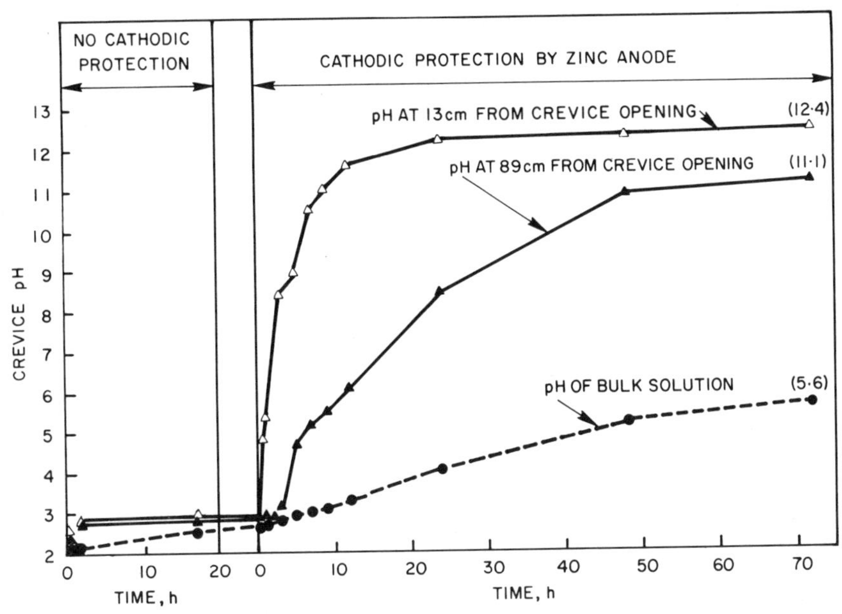

Figure 5-11. Change of crevice pH with time in a type 304 crevice connected to a zinc anode in an acidified 0.6 *M* NaCl solution. (After Lennox and Peterson.)

seawater. It is not surprising, therefore, that no simple relationship has been found between temperature and crevice attack, as demonstrated by the data given in Table 5-10.

Regarding the effect of velocity of the corrodent on crevice corrosion, studies by Peterson et al. (5) of corrosive attack at type 304 O-ring seals suggest that seawater flowing at 0.15 m/second generates more crevice attack than quiescent seawater. Such behavior would be anticipated if it is assumed that the moving water provides a greater supply of cathodically reducible oxygen to the surface outside the crevice. However, if the crevice is formed between marine growths or deposits and the stainless steel, then increasing the velocity of the seawater may be beneficial in that growths or deposits do not adhere in high velocity seawater.* There is also some evidence to suggest that increasing seawater velocity to 1.2 m/second decreases crevice attack in mechanical crevices (30).

The effect of surface condition on crevice corrosion of stainless steels in seawater has been evaluated by Degerbeck (24) and others (31). The former study revealed that ground surfaces exhibited more crevice attack

*If flow is interrupted enabling growths to attach, restoring flow will not remove them.

TABLE 5-10. EFFECT OF SEAWATER TEMPERATURE ON THE PROBABILITY OF CREVICE CORROSION INITIATION. MULTIPLE CREVICE TEST, 30 DAY EXPOSURE

Stainless Steel	Probability of Crevice Corrosion Initiation, %			
	5°C	15°C	25°C	50°C
Type 304	2	13	70	90
Type 316	0	2	51	21
Type 317	0	0	1.4	4
Type 430	0.8	52	98	98
Type 446	0	7	43	19
E-BRITE 26-1	0	0	0	4

than pickled surfaces. Other investigations (31) have been less conclusive in identifying effects of surface finish.

5-6 AVOIDANCE OF CREVICE CORROSION

Concerning measures that can be taken to minimize crevice corrosion, Flint (32) has noted the following:

1. Design to avoid formation of a crevice or at least to keep it as open as possible.
2. Overlay susceptible areas with a more crevice corrosion resistant alloy.
3. Paint surrounding cathodic surfaces.
4. Use inhibiting paste.*
5. Avoid metal/nonmetal joints since these are usually tighter than metal/metal joints.

In addition to these measures, cathodic protection and increasing velocity could also be tried in some instances. However, the present state-of-the-art suggests that to ensure freedom from crevice corrosion, it may be prudent to utilize high molybdenum stainless steels or higher alloys containing molybdenum for components that are to operate under conditions in which crevice corrosion is a possibility.

REFERENCES

1. I. L. Rosenfeld, *Localized Corrosion*, National Association of Corrosion Engineers, Houston, Tex., 1974, p. 373.
2. F. L. LaQue, *Marine Corrosion—Causes and Prevention,* John Wiley & Sons, New York, 1975, p. 164.

*For results see publication by E. H. Wyche et al., *Trans. Electrochem. Soc.* Ch. 7, Vol. 89, p. 149, 1946.

3. W. D. France, "Crevice Corrosion of Metals," *Localized Corrosion—Cause of Metal Failure,* ASTM-STP516, American Society for Testing and Materials, 1972, p. 164.
4. M. G. Fontana and N. D. Greene, *Corrosion Engineering,* McGraw-Hill Book Company, New York, 1967.
5. M. H. Peterson, T. J. Lennox, and R. E. Groover, *Mater. Prot.*, p. 23, January 1970.
6. F. D. Bogar and C. T. Fujii, *N. R. L. Report 7690,* Naval Research Laboratory, Washington, D. C., 1974.
7. N. D. Greene, W. D. France, and B. E. Wilde, *Corrosion,* Vol. 21, No. 9, p. 275, 1965.
8. D. Sinigaglia, D. Taccani, and G. Bombara, *Electrochim. Met.*, Vol. 3, p. 297, 1968.
9. B. E. Wilde, *Localized Corrosion,* National Association of Corrosion Engineers, Houston, Tex., 1974, p. 342.
10. J. C. Rowlands, *Br. Corros. J.*, Vol. 11, No. 4, p. 195, 1976.
11. W. D. France and N. D. Greene, *Corrosion,* Vol. 20, p. 247, 1968.
12. G. Bombara, D. Sinigaglia, and D. Taccani, *Electrochim. Met.*, Vol. 3, p. 81, 1968.
13. J. W. Oldfield and W. H. Sutton, *Crevice Corrosion of Stainless Steels,* paper presented at the 6th European Congress on Corrosion, London, September 1977.
14. D. B. Anderson, "Statistical Aspects of Crevice Corrosion in Sea Water," *Galvanic and Pitting Corrosion—Field and Laboratory Studies,* ASTM-STP576, American Society for Testing and Materials, 1976, p. 231.
15. *Annual Book of ASTM Standards,* Amercian Society for Testing and Materials, Philadelphia, Pa., 1976.
16. T. P. May and H. A. Humble, *Corrosion,* Vol. 5, No. 2, p. 50, 1952.
17. K. D. Efird, *The Effect of Temperature on Crevice Corrosion of Stainless Steels in Seawater,* paper presented at the 6th International Congress on Metallic Corrosion, December 1975, Sydney, Australia.
18. E. A. Lizlovs, "Crevice Corrosion of Some High Purity Ferritic Stainless Steels," *Localized Corrosion—Cause of Metal Failure*, ASTM STP516, American Society for Testing and Materials, 1972, p. 201.
19. E. C. Hoxie, "Some Corrosion Considerations in the Selection of Stainless Steel for Pressure Vessels and Piping," in *Pressure Vessels and Piping: Decade of Progress,* Vol. 3, The American Society of Mechanical Engineers, New York, 1977.
20. R. J. Brigham, *Mater. Performance,* Vol. 13, p. 29, November 1974.
21. R. P. Jackson and D. van Rooyen, "Crevice Corrosion of Some Ni-Cr-Mo-Fe Alloys in Laboratory Tests," *Localized Corrosion—Cause of Metal Failure,* ASTM-STP516, American Society for Testing and Materials, 1972, p. 210.
22. J. R. Ambrose, in *Passivity of Metals*, The Electrochemical Society, Princeton, N. J., 1978, p. 740.
23. S. Berg and S. Henriksson, *J. Iron Steel Inst.,* Vol. 199, p. 188, 1961.
24. J. Degerbeck, *Chem. Process Eng.*, p. 47, December 1971.
25. M. A. Streicher, *Platinum Met. Rev.,* Vol. 21, No. 2, p. 51, 1977.
26. O. B. Ellis and F. L. LaQue, *Corrosion,* Vol. 7, No. 11, p. 362, 1951.
27. T. P. May and H. A. Humble, *Corrosion,* Vol. 8, No. 2, p. 50, 1952.
28. T. J. Lennox and M. H. Peterson, *N. R. L. Report 2374,* Naval Research Laboratory, Washington, D. C., 1971.
29. R. E. Groover and M. H. Peterson, *Mater. Performance,* p. 24, November 1974.
30. G. E. Moller, *J. Soc. Pet. Eng.*, p. 101, April 1977.
31. *Metals Handbook*, 8th ed., Vol. 1, American Society for Metals, Metals Park, Ohio, 1969, p. 560.
32. G. N. Flint, *Resistance of Stainless Steels to Corrosion in Naturally Occurring Waters,* paper presented at 2nd Spanish Corrosion Congress, Zaragoza, Spain, May 1976.

6

INTERGRANULAR CORROSION

6-1 INTRODUCTION

Grain boundaries are disordered misfit regions separating grains of different crystallographic orientation, and therefore they are favored sites for the segregation of various solute elements or the precipitation of metal compounds, such as carbides and sigma. It is not surprising, therefore, that in certain corrodents grain boundaries can be preferentially attacked. This type of attack is known as intergranular corrosion and most metals and alloys in specific corrodents can exhibit it.

In stainless steels intergranular corrosion can occur as a result of segregation of certain solute elements to grain boundaries. However, this is known to occur only in highly oxidizing corrodents containing cations in high valence states, such as the hexavalent chromium ion (Cr^{+6}), and is generally regarded as a somewhat unusual case of intergranular corrosion not encountered in service. The usually encountered form of intergranular corrosion is due to sensitization, an example of which is shown in Figure 6-1. As discussed in Chapter 2, sensitization is now fairly well understood in terms of chromium carbide precipitation, and is attributed to chromium depletion in the vicinity of carbides precipitated at grain boundaries.

It should be emphasized, however, that even if a given stainless steel is in the sensitized condition, it will not necessarily exhibit intergranular attack in all service environments. For example, in a study reported by the Welding Research Council (1) in which nine different austenitic stainless steels were tested in the sensitized condition in many industrial environments, 35 environments produced intergranular attack and 39 did not.

Some attempt at seeking to understand such variability can be made by considering the active-passive behavior of stainless steels defined in Fig-

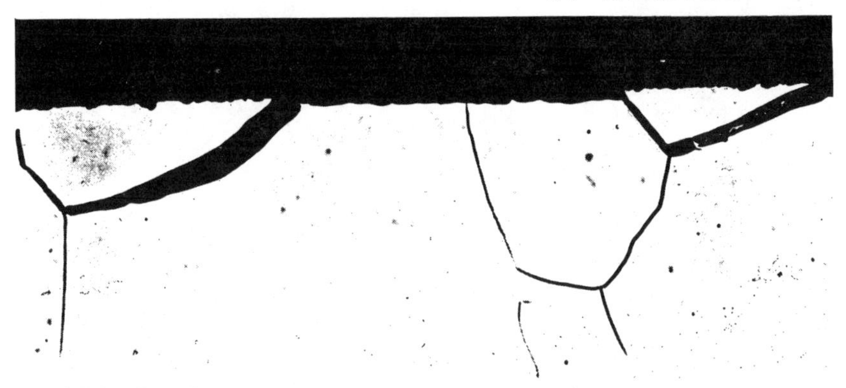

Figure 6-1. Intergranular attack in a sensitized austenitic alloy produced by exposure to a boiling sulfuric acid-ferric sulfate solution. Prolonged exposure causes grains to detach from surface. (100×)

ure 3-5. Clearly in environments where intergranular corrosion takes place the sensitized boundary is exhibiting active behavior while the grains exhibit passive behavior. Conversely, in environments where intergranular corrosion does not take place both the sensitized boundary and the grains exhibit passive behavior. Some insight into the changes of the active-passive behavior of high chromium grains and low chromium grain boundary regions can be obtained from the anodic polarization studies of Osozawa and Engell (2), who evaluated a number of Fe-Ni-Cr alloys of varying chromium content in a sulfuric acid solution (Figure 6-2). It can be seen that as the chromium content of the alloy is reduced to below approximately 12%, there is a marked decrease in the width of the passive potential range (defined in Figure 3-5). Thus, if sensitization reduces the chromium content at the grain boundaries to values below 12% while the grains retain a chromium content of about 18%, there will be a significant potential range over which the boundaries will exhibit active corrosion while the grains remain passive. There will also be potential ranges over which both the high chromium grains and the low chromium boundary regions will exhibit passive behavior. Therefore, conceptually at least, it should be possible to define sensitized stainless steel/environment combinations in which intergranular attack is or is not likely to occur as a result of chromium depletion by considering differences in passive potential ranges. However, historically intergranular corrosion was recognized and studied using certain acid immersion tests well before electrochemical techniques were developed. These acid tests have remained in use as quality control or acceptance tests in industries

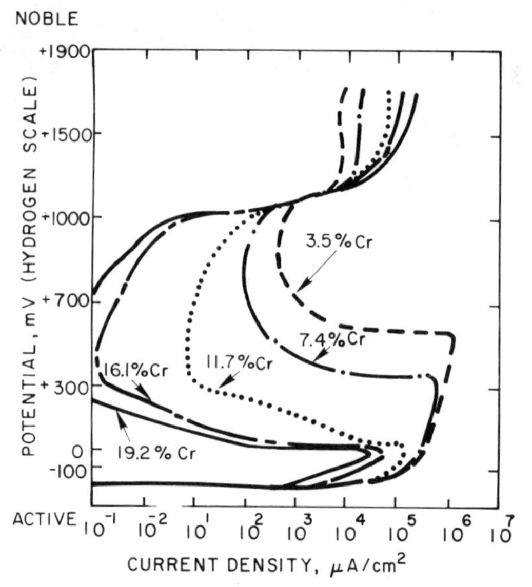

Figure 6-2. Effect of chromium content of Fe-Ni-Cr alloys on their anodic polarization behavior in a 2 *N* H_2SO_4 solution at 90°C. Nickel content in the range 8.3–9.8%. (After Osozawa and Engell.)

using stainless steels primarily because of extensive existing correlations with service experience.

6-2 TECHNIQUES FOR MEASURING TENDENCY TO INTERGRANULAR CORROSION

6-2-1 Acid Tests

The acid tests standardized by the ASTM for evaluating resistance to intergranular corrosion are listed in Table 6-1. These test media may bear little relationship to the intended service environment, but are believed by their users to be capable of detecting in relatively short periods of time metallurgical conditions either suspected or known to cause intergranular attack in some service environments. The ASTM standard tests listed in Table 6-1 were developed for austenitic stainless steels. Standardized testing procedures for ferritic stainless steels are still being considered by the ASTM.

The copper sulfate-sulfuric acid test (ASTM A393-63), commonly known as the Strauss test, was the earliest test used to detect susceptibility to intergranular corrosion in sensitized stainless steels (3, 4). The test specimen was exposed to the testing solution for 72 hours, and damage

TABLE 6-1. STANDARD INTERGRANULAR CORROSION TESTS FOR STAINLESS STEELS

ASTM Standard (Common Name)	Environment	Exposure	Evaluation	Species Attacked
A393-63* (Strauss)	15.7%H_2SO_4 + 5.7%$CuSO_4$, Boiling	One 72 h Period	Appearance After Bending	Chromium Depleted Area
A262-70, Practice A (Oxalic Etch)	10%$H_2C_2O_4$, Anodic at 1 A/cm^2, Ambient Temperature	1.5 min.	Type of Attack	Various Carbides
A262-70, Practice B (Streicher)	50%H_2SO_4+2.5%$Fe_2(SO_4)_3$, Boiling	One 120 h Period	Weight Loss Per Unit Area	Chromium Depleted Area
A262-70, Practice C (Huey)	65%HNO_3, Boiling	Five 48 h Periods, Fresh Solution Each Period	Average Weight Loss Per Unit Area	Chromium Depleted Area, Sigma and Carbides
A262-70, Practice D (Warren)	10%HNO_3 + 3%HF, 70°C	Two 2 h Periods	Weight Loss Per Unit Area	Chromium Depleted Area in Mo-Bearing Steels
A262-70, Practice E (Copper Accelerated Strauss)	15.7%H_2SO_4 + 5.7$CuSO_4$, Specimen in Contact with Copper, Boiling	One 24 h Period	Appearance After Bending	Chromium Depleted Area

*Discontinued in 1972.

assessed by bending through 180° and examining the outside surface for accentuated intergranular penetrations. While this test was adequate for the high carbon stainless steels produced in earlier days, it is not considered sufficiently severe for modern lower carbon stainless steels. This test was modified (5, 6) to incorporate contact with metallic copper during exposure which increases the severity of attack (ASTM A262-70, Practice E). As in the earlier version, specimens are classified as acceptable or unacceptable depending on whether cracks are seen in bent specimens.

The oxalic acid test (ASTM A262-70, Practice A) developed by Streicher (7) is a rapid etching procedure used to screen out specimens that would unquestionably pass the more lengthy and costly boiling nitric acid test (ASTM A262-70, Practice C). The etched structures are classified as exhibiting "step" (free of carbide), "ditch" (continuous carbide), and "dual" (intermediate) modes of attack.

The ferric sulfate-sulfuric acid test (ASTM A262-70, Practice B), originally described by Streicher (8), is of comparable severity to the boiling nitric acid test and has several advantages over the latter. For example, exposure times are shorter, as demonstrated in Table 6-1, an as yet unidentified phase formed in types 316L and 317L by sensitizing treatments [known as submicroscopic sigma (9)] is not attacked, and accumulation of corrosion products in the solution does not accelerate corrosion rate. However, attack attributed to "submicroscopic sigma" may occur in the stabilized grades, types 321 and 347.

The boiling nitric acid test (ASTM A262-70, Practice C), commonly known as the Huey test (10), has been the most popular of all the intergranular corrosion tests in the U. S. Details of this test are shown in Table 6-1. Yet there are several disadvantages to this test. As pointed out by Henthorne (11) the three main disadvantages are (1) the test takes a long time and is expensive, (2) intergranular attack is accelerated by the hexavalent chromium ion whose concentration builds up during each 48 hour test period, and (3) the test is sensitive to surface preparation, end grain structure, and attacks phases other than chromium carbide, namely, sigma, "submicroscopic sigma," and titanium carbide.

The nitric acid-hydrofluoric acid test (ASTM A262-70, Practice D) was developed by Warren (12) to differentiate between the effects of chromium depletion and sigma in molybdenum containing types 316, 316L, 317, and 317L. The test solution does not attack sigma. However, because the general corrosion rate in this solution is quite high, it is necessary to make careful comparisons between weight losses of sensitized and nonsensitized material.

There is a considerable amount of published information regarding the evaluation of stainless steels by these acid tests. For more detail the

TABLE 6-2. MAXIMUM ACCEPTABLE EVALUATION TEST RATES SPECIFIED BY duPONT FOR SERVICES WHERE SUSCEPTIBLE MATERIAL WOULD SUFFER INTERGRANULAR ATTACK

Type	Condition	Maximum Corrosion Rate, inches per month
120 HOUR FERRIC SULFATE-SULFURIC ACID TEST (ASTM A-262, PRACTICE B)		
304	As-Received	0.0040
304L	20 Min, 677°C	0.0040
316	As-Received	0.0040
316L	20 Min, 677°C	0.0040
317L	20 Min, 677°C	0.0040
CF-8	As-Received	0.0040
CF-8M	As-Received	0.0040
240 HOUR NITRIC ACID TEST (ASTM A-262, PRACTICE C)		
304	As-Received	0.0015
304L	20 Min., 677°C	0.0010
304L	1 h, 677°C	0.0020
309S	As-Received	0.0010
316	As-Received	0.0015
347	1 h, 677°C	0.0020
CF-8	As-Received	0.0020
CF-8M	As-Received	0.0025

reader is referred to the reviews by Henthorne (11), Brown (9), and Cowan and Tedmon (13).

It should be noted that the various practices of ASTM A262-70 specify only details of testing and do not comment on what are acceptable rates of integranular corrosion. The maximum corrosion rates reported to be acceptable by duPont (9) are shown in Table 6-2. These rates are expressed in inches per month, having been determined by weight loss methods using the ferric sulfate-sulfuric acid and nitric acid tests. It is claimed that these acceptance rates were derived by correlating thousands of laboratory tests with many years of service experience (9). It has also been noted that Du Pont specifies intergranular corrosion tests for only about 15% of the stainless steels purchased (14).

In order to describe the effects of varying elevated temperatures and holding times on subsequent intergranular corrosion behavior, it is sometimes convenient to present data in terms of a time-temperature-sensitization (TTS) diagram. An example of a TTS diagram is shown in Figure 6-3 (15). These diagrams are established by determining rates of attack in the various acid media noted after heating a given material to various temperatures and holding at temperature for various times. Contours are drawn to separate time-temperature combinations yielding certain rates of attack.

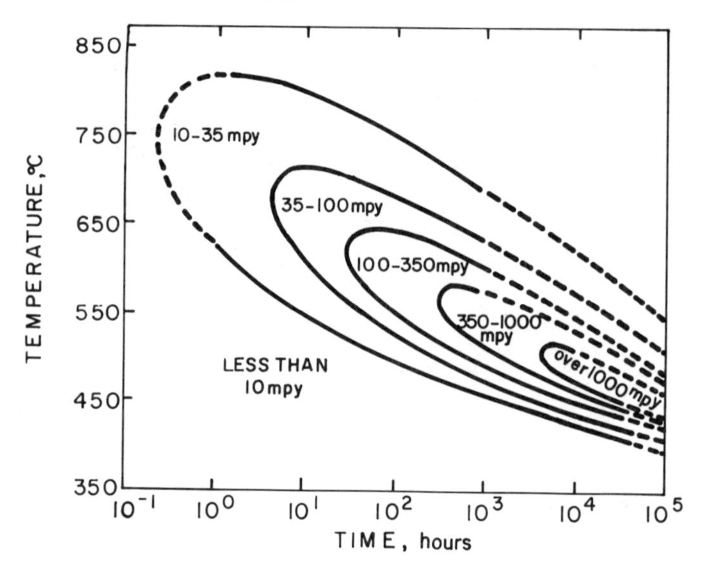

Figure 6-3. Time-temperature-sensitization curves for type 347 stainless steel obtained by the Huey test. (After Ebling and Scheil.)

6-2-2 Electrochemical Evaluations

As noted in Section 6-1, no electrochemical techniques for determining intergranular attack have yet been standardized by the ASTM. Nevertheless, numerous articles have appeared in the literature describing procedures for studying susceptibility to intergranular attack by such techniques. For a review and discussion of these studies the reader should consult the publications by Henthorne (11) and Cowan and Tedmon (13).

In principle there is no reason why electrochemical techniques should not be used in evaluating susceptibility to intergranular corrosion. Present understanding suggests that intergranular corrosion is primarily under anodic control (i.e., attack is determined by the availability of anodic sites at the grain boundaries). Therefore, it would be expected that the anodic polarization curve of a sensitized stainless steel would be different from that of a nonsensitized material. This is in fact the case, as demonstrated by the measurements of Osozawa et al. (16) which are given in Figure 6-4. Among parameters that are particularly affected by sensitization are i_{pass} and i_{crit} defined in Figure 3-5. However, it is also evident from Figure 6-4 that one of the largest measurable changes is in the area defining the "anodic nose" corresponding to the active-passive transition. Recently, Clarke, Romero, and Danko (17) have studied the possibility of using the measurement of this area as a basis for assessing the degree of sensitiza-

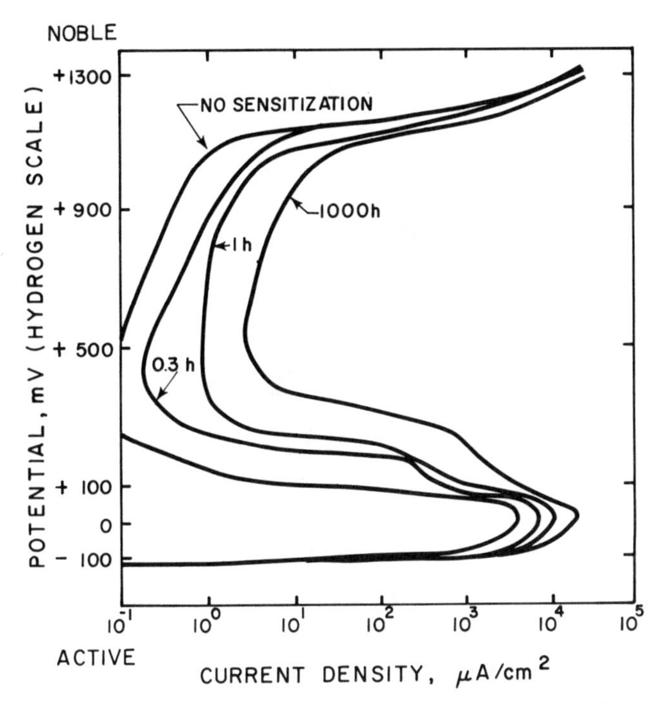

Figure 6-4. Anodic polarization curves of type 304 stainless steel in a 2 *N* H_2SO_4 solution at 90°C after sensitization at 650°C for various times. (After Osozawa, Bohnenkamp, and Engell.)

tion of type 304 stainless steel. This study was prompted by the desirability of having a rapid, nondestructive field test for determining the degree of sensitization in field welded components for the nuclear power industry. It is not difficult to envision the usefulness of such a test also in other industries. It seems possible, therefore, that the development of electrochemical techniques for detecting sensitization in stainless steels may progress in the area of field testing of welded components, where the standard acid immersion tests obviously cannot be used.

6-3 SENSITIZATION PRODUCED BY WELDING

Since welding is a common method of joining stainless steels for corrosion resistant service, it is important to be able to determine the extent to which welding can cause sensitization and hence intergranular attack. Such knowledge can be useful in comparing alloys and welding processes, as well as in quality control and in the development of new alloys. As noted in Chapter 2, the formation of chromium depleted zones at grain

TABLE 6-3. FACTORS THAT CAN INFLUENCE WELDMENT CORROSION

1. Composition and structure of base metal and weld metal.
2. Metallurgical condition of base metal before welding (prior thermal and mechanical working history).
3. Welding process - TIG, MIG, shielded metal-arc and submerged-arc.
4. Welding procedure - manual or automatic, number of passes, rate of welding, current and voltage.
5. Shielding gas - composition and flow rate.
6. Size of material welded (i.e., plate thickness) and size and geometry of weld deposit.

boundaries requires very specific combinations of temperature, time at temperature, and composition. In material adjacent to welds (i.e., within the heat affected zone) such combinations may or may not occur depending on a number of factors. The various factors that can influence weldment corrosion are listed in Table 6-3 (18). As discussed in Chapter 2, for austenitic stainless steels the use of low carbon grades, stabilized grades, or postwelding heat treatment can be effective expedients in minimizing intergranular attack at welds.

The current state-of-the-art of testing weldments to determine resistance to intergranular attack has been covered in a relatively recent publication by Henthorne (18). For austenitic and ferritic stainless steels, all the ASTM practices listed in Table 6-1 have been used in the past. For higher alloys, specifically Hastelloy alloys C, C-276 and G, Carpenter 20Cb-3, Inconel alloys 600 and 625, and Incoloy alloys 800 and 825, the ferric sulfate-sulfuric acid test has been recommended as described in ASTM G28-72.

As pointed out by Henthorne (18), posttest evaluations of welds employing weight loss could be misleading, since a sensitized region within the heat affected zone at which attack is occurring can often represent a small fraction of the test specimen and hence yield low overall weight losses. Accordingly, microscopic examination for evidence of intergranular attack is recommended whenever possible.

Reservations have also been expressed about the use of heat treatments to simulate welding effects on the grounds that such practices may be too conservative. Table 6-4 (1) is a summary of the poor correlation observed

TABLE 6-4. COMPARATIVE SEVERITY OF WELDING AND HEAT TREATMENT ON INTERGRANULAR CORROSION

AISI Type	Number of Racks Showing Intergranular Attack[a]	
	Welded	Heat Treated[b]
316L	0	11
316	1	15
317	6	22
302	15	19
304	1	10
304L	0	5
321	6	12
347	1	13

[a]Racks (24 specimens) exposed to a number of actual service environments for times ranging from 30 to 1600 days.
[b]Specimens heat treated for 1 to 4 hours at 593 to 677°C.

in the susceptibility to intergranular attack between welded and furnace sensitized austenitic stainless steels. While no such comparisons exist for ferritic stainless steels, it seems also unlikely that for these materials heat treatments will give a good indication of the behavior of welds, particularly since cooling rate may be very important in determining whether sensitization occurs. It would seem, therefore, that actual weldments should be tested whenever possible. Nevertheless, it is difficult to criticize conservative evaluation practices for intergranular corrosion testing, since in certain segments of the chemical process industry equipment failures by intergranular corrosion cannot be considered insignificant, as demonstrated in Table 1-3.

6-4 KNIFE-LINE ATTACK

As described in Chapter 2, one of the metallurgical remedies to counter sensitization of austenitic stainless steels is to alloy with strong carbide formers, such as titanium or columbium. The grade alloyed with titanium is type 321 and the grade alloyed with columbium (plus tantalum) is type 347 (see Figure 2-1). These grades are commonly referred to as "stabilized." The term stabilized implies that little carbon remains in solid solution to be precipitated as chromium carbide, having been precipitated as titanium or columbium carbide at higher temperatures. Titanium and columbium carbides form in the temperature range 870–1150°C, whereas chromium carbides form in the range 480–760°C. Thus types 321 and 347, as received from the steel mill, have most of their carbon precipitated as titanium or columbium carbides, having been "mill annealed'' at about 1065°C and rapidly cooled to room temperature. When such mill annealed types 321 and 347 are welded, the temperature in a very narrow zone immediately adjacent to the weld more than exceeds 1150°C and redissolves the titanium or columbium carbides, putting carbon back into solid solution within this narrow zone. If the subsequent cooling rate is very slow through the range 480–760°C, or if the weld is reheated to temperatures in the chromium carbide precipitation range (480–760°C), a very thin zone of sensitized material may develop immediately adjacent to the weld that can be susceptible to intergranular attack in certain environments. Because the sensitized region is very narrow, attack, if it occurs, appears to be along a thin line right along the weld/base metal junction. It has therefore been termed ''knife-line attack'' (19, 20). Knife-line attack can be eliminated by reheating the welded part to about 1065°C to redissolve chromium carbide and reprecipitate titanium or columbium carbides.

More recent studies by Cihal (21) have also emphasized the importnce of the morphology of the carbide precipitates at grain boundaries in determining whether knife-line attack will occur. Furthermore, it is not known whether knife-line attack is confined solely to the chromium depleted regions in the vicinity of chromium carbides. Knife-line attack is found mostly in oxidizing environments, such as nitric acid. Therefore, it is possible that in addition to chromium depleted regions, the carbides themselves and any sigma present may also be attacked, since nitric acid is known to attack these phases (see Huey test, Table 6-1). In addition to nitric acid, certain oxidizing urea environments are reported to cause knife-line attack (21). The use of low carbon grades, such as type 304L, has given superior performance in environments where the stabilized grades have exhibited knife-line attack (22).

6-5 POLYTHIONIC ACID CRACKING

The instances of intergranular corrosion discussed so far do not require the presence of an externally applied stress. However, in the 1950s a form of intergranular corrosion requiring the presence of stress became recognized as a technological problem in catalytic reformers used in the petroleum industry. Studies by Dravnieks and Samans (23, 24) showed that this form of attack occurred in the aqueous condensates of reforming furnaces when polythionic acids ($H_2S_xO_6$, where $x = 3$, 4, or 5) were present. Subsequently this form of attack became known as polythionic acid cracking. However, it is by no means clear that the presence of the polythionic acid species is essential, since similar attack can occur in water saturated with sulfur dioxide (25, 26). However, because of its common usage, the term polythionic acid cracking is used here.

A laboratory test for resistance to polythionic acid cracking has been standardized as ASTM Practice G 35-73. The solution is prepared by bubbling sulfur dioxide and then hydrogen sulfide through distilled water. Stressed U-bend specimens are often used in such tests, although any type of stress corrosion test specimen can be used. Samans (24) has shown an excellent correlation between polythionic acid cracking resistance as determined by ASTM G 35-73 and intergranular corrosion resistance as determined by the Strauss test (ASTM A393-63, described in Table 6-1), for a dozen different austenitic stainless steels and higher alloys. An example of this correlation in terms of a TTS diagram is shown in Figure 6-5. As a result of this study, it is now generally accepted that polythionic acid cracking is a form of intergranular corrosion which occurs in the presence of applied stress. Accordingly, attempts to mitigate this form of attack have included the metallurgical remedies used to minimize intergranular corrosion.

The technological problem encountered in the petroleum industry is essentially as follows. Stainless steel components are used at operating temperatures that may be high enough to sensitize the material. On subsequent shutdown of a refinery unit, polythionic acids may form by the interaction of sulfur compounds, moisture, and air at low temperatures, and attack the chromium depleted boundaries. A combination of operational procedures and metallurgical remedies have been used to minimize polythionic acid cracking during shutdowns. The operational procedures

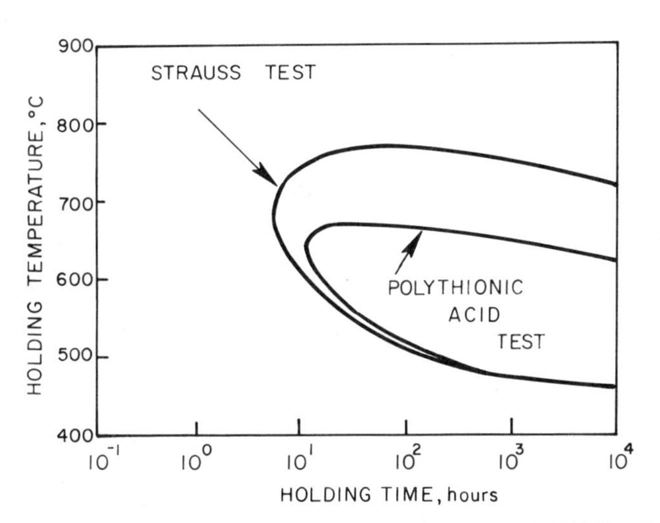

Figure 6-5. Effect of heat treatment on the resistance of type 304 (0.04% C) stainless steel in polythionic acid and Strauss tests. (After Samans.)

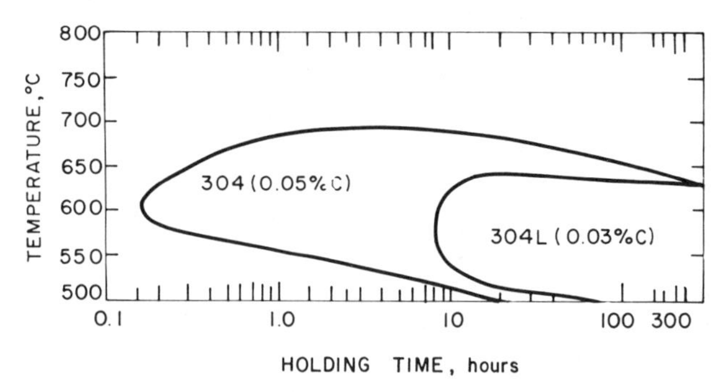

Figure 6-6. Time-temperature-sensitization curves for type 304 and 304L stainless steels obtained using the Strauss test. (Courtesy of Sandvikens Jernverks Aktiebolag.)

are described in the NACE Recommended Practice RP-01-70, and are based on preventing the condensation of water vapor or maintaining an alkaline environment by additions of ammonia or sodium carbonate.

Among the metallurgical remedies that have been tried are the use of low carbon grades (e.g., type 304L), the use of stabilized grades (types 321 or 347), and the use of higher alloys such as Incoloy alloy 801. As shown in Figures 6-3 and 6-6, the stabilized and the low carbon grades of stainless steel can be sensitized at certain time-temperture combinations. Accordingly, it has been recommended that low carbon grades be used for operating temperatures below 400°C, and that stabilized grades be thermally treated at temperatures of 900–925°C to ensure maximum precipitation of carbon as titanium or columbium carbides prior to service (27). In cases where welding is necessary, a British practice has been to weld the thermally stabilized (920°C) type 321 and to locally anneal the welds at 920°C (28). Higher alloys, such as the titanium stabilized Incoloy alloy 801, have been used in the U. S. As shown by Stephens and Scarberry (29), if alloy 801 is thermally stabilized at 955°C, subsequent heat treatments do not produce the degree of sensitization required for polythionic acid cracking, as given in Table 6-5.

6-6 EFFECT OF COMPOSITION AND STRUCTURE

6-6-1 Austenitic Stainless Steels

As discussed in Chapter 2, the key alloying element responsible for sensitization is carbon. The effects of the other alloying elements on sensitization are related primarily to their influence on the solubility and precipitation of carbides. The importance of carbon on intergranular corrosion in the Huey test is illustrated by Figure 6-7 (30), which also

Table 6-5. Effect of Heat Treatment on the Corrosion Rate in the Huey Test and on the Resistance to Polythionic Acid Cracking of Incoloy Alloys 800 and 801

Heat Treatment	Alloy 800[a]		Alloy 801[b]	
	Corrosion Rate in Boiling 65% HNO_3, mpy	Time to Cracking in Polythionic Acid, h	Corrosion Rate in Boiling 65% HNO_3, mpy	Time to Cracking in Polythionic Acid, h
955°C/1 h/WQ (=A)	6.0	NC	--	NC
A + 650°C/1 h/AC	1200	48	12.0	NC
A + 650°C/2 h/AC	1200	7	10.8	NC
A + 650°C/4 h/AC	808	NC	15.6	NC
A + 705°C/4 h/AC	16.8	NC	9.6	NC
A + 815°C/4 h/AC	12.0	NC	12.0	NC
A + 870°C/4 h/AC	8.4	NC	26.4	NC

WQ = water quenched.
AC = air cooled.
NC = no cracking in 1000 h test.

[a] Nominal composition, 46%Fe-32.5%Ni-21%Cr-0.05%C.
[b] Nominal composition, 44.5%Fe-32%Ni-20.5%Cr-0.05%C-1.1%Ti.

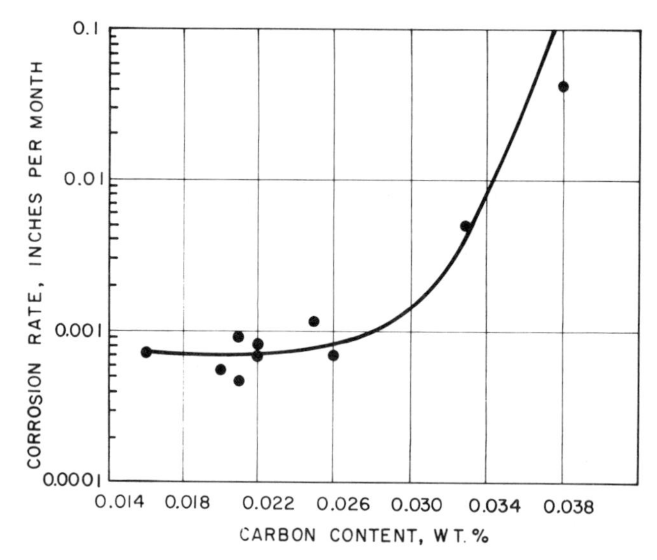

Figure 6-7. Effect of carbon content on the corrosion rate in the Huey test of type 304 stainless steel heated for two hours at 650°C. (After Heger and Hamilton.)

shows the reason for adopting the 0.03% C (maximum) specification for the low carbon (L) grades. At carbon contents greater than 0.03% there is a rapid increase in corrosion rate.

The combinations of chromium, nickel, and carbon that must be maintained to prevent intergranular attack of sensitized materials in the Strauss test are illustrated in Figure 6-8 (31). Essentially, chromium increases and nickel decreases the solubility of carbon in austenite, so a balance must be struck between the contents of these three elements. However, this pattern cannot be extrapolated directly to higher nickel alloys. For example, Inconel alloy 690 (60% Ni-30% Cr-10% Fe) with a carbon content of 0.02%, which is a carbon level readily attainable with commercial AOD melt practice, cannot be sensitized to show any intergranular attack in the severe Huey test, as illustrated by Figure 6-9 (32).

Molybdenum, at the 2 to 3% level present in type 316 stainless steel was regarded in the past as detrimental to intergranular corrosion resistance. For example, in the Huey test high rates of intergranular attack are observed in sensitized type 316 even at carbon contents below the maximum specified for the L-grade as shown in Figure 6-10 (30). Current thinking is that these high rates of attack are peculiar to the Huey test, which attacks a number of phases in addition to chromium depleted regions (see Table 6-1). In the Strauss test, the presence of 2% molyb-

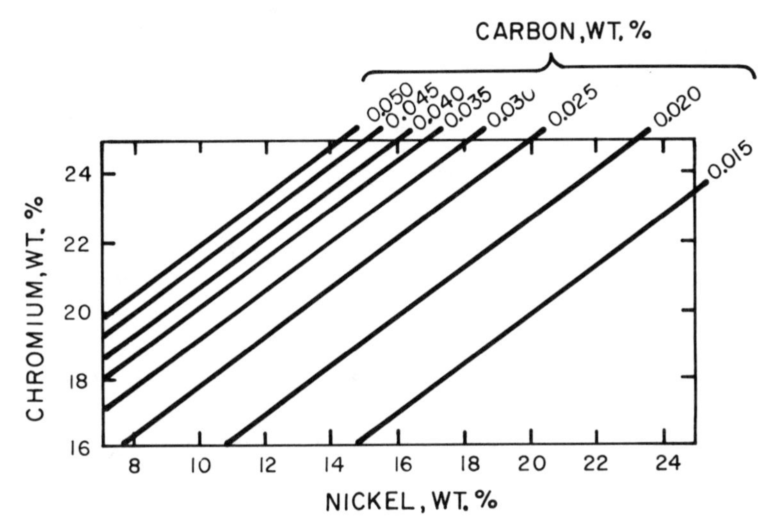

Figure 6-8. Effect of chromium and nickel content on the carbon content required to avoid intergranular attack in stainless steels in the Strauss test after sensitization at 650°C for one hour. (After Cihal.)

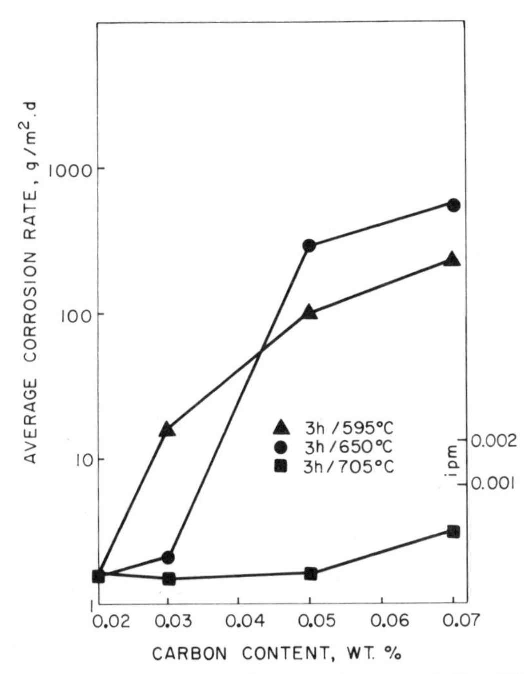

Figure 6-9. Effect of carbon content on the corrosion rate of alloy 690 in the Huey test. Materials annealed for one hour at 1150°C and water quenched before given sensitizing heat treatments. (After Sedriks, Schultz, and Cordovi.)

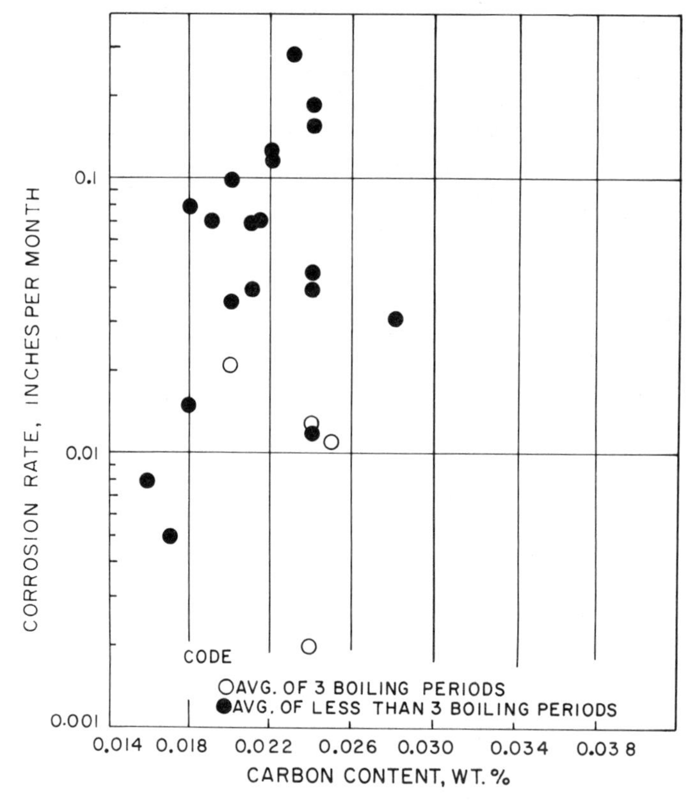

Figure 6-10. Effect of carbon content on the corrosion rate in the Huey test of type 316 stainless steel sensitized at 675°C for two hours. (After Heger and Hamilton.)

denum actually reduces the tendency to sentitization, as illustrated in Figure 6-11 (33). The reason for the high rates of attack observed in type 316L in the Huey test has never been conclusively established. Since molybdenum favors the formation of the sigma phase, this effect has been attributed to the presence of "submicroscopic sigma." This description is further endorsed by the fact that no sigma phase has ever been detected by microscopic techniques (34). Warren (35) has indicated that aside from the Huey test, submicroscopic sigma is known to have caused intergranular attack only in one other environment, namely 25% H_2SO_4 at 40°C. Nevertheless, from a practical viewpoint, it would seem prudent to test molybdenum containing grades in any new sulfuric acid or nitric acid application before specifying their use. Also, further research seems desirable to identify the exact nature of the attack attributed to submicro-

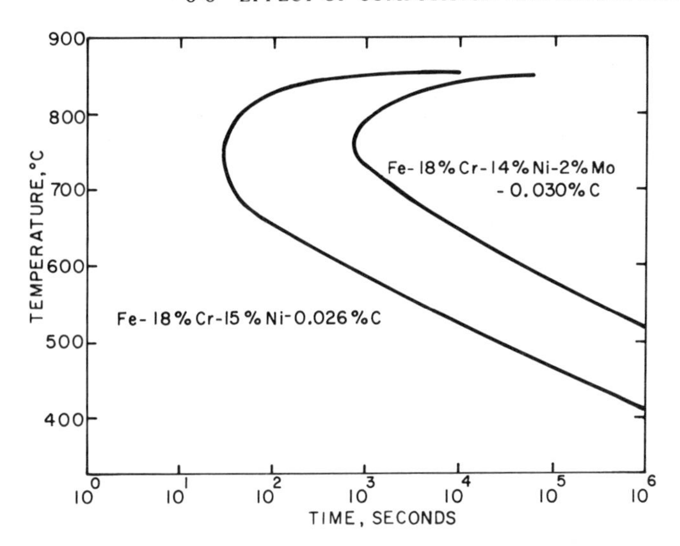

Figure 6-11. Effect of molybdenum content on the sensitization behavior of austenitic stainless steels in the Strauss test. (After Binder, Brown, and Franks.)

scopic sigma. The possibility that this effect is due to segregation of some solute elements rather than the precipitation of a distinct phase cannot be ruled out.

Sigma, when present as an identifiable phase at grain boundaries, is attacked in the Huey test (see Table 6-1) and hence should be avoided in stainless steels used for nitric acid service. However, the molybdenum containing types 316L and 317L are not usually selected for nitric acid service because type 304L exhibits better corrosion resistance in nitric acid and is less expensive.

Titanium carbide, present in titanium stabilized type 321, can also be attacked in certain strongly oxidizing acids such as nitric acid (36, 37), and if precipitated at grain boundaries could lead to intergranular attack in oxidizing environments. This is one of the reasons why stabilization with columbium (type 347) is sometimes preferred. Steelmakers prefer columbium because it can be added in the melting furnace, allowing time for analysis and corrections, whereas titanium, because of its higher reactivity, has to be added to the ladle during pouring.

High manganese contents, such as are found in the AISI 200 series of stainless steels, produce a slightly detrimental effect on resistance to intergranular attack as noted by the comparison with type 304 given in Table 6-6 (38). The higher manganese type 202 appears to be less resistant than the lower manganese type 201 in the sensitized condition.

TABLE 6-6. RESULTS OF INTERGRANULAR CORROSION TESTS ON NICKEL-MANGANESE-NITROGEN STAINLESS STEELS

AISI Type	Corrosion Rate in Huey Test, inches per month	Result of Strauss Test
	AS-RECEIVED (MILL ANNEALED)	
201	0.0015[a]	Passed
202	0.0021[a]	Passed
304	0.0013[a]	Passed
	SENSITIZED, 1 h/650°C/AC	
201	0.0807[b]	Failed
202	0.2225[b]	Failed
304	0.0657[b]	Failed

[a]Average of five 48 h test periods.
[b]One 48 h test period.

Silicon has been shown to decrease resistance to intergranular corrosion (39–41), especially in molybdenum containing stainless steels (42). However, as noted at the end of this chapter, high levels of silicon prevent attack in nonsensitized stainless steels by nitric acid solutions containing the hexavalent chromium ion.

Boron has been reported to cause both a beneficial and deleterious effect on intergranular attack (11) and further studies are needed to clarify its effect.

The effect of nitrogen on sensitization is complex and depends on the amount present in the alloy and the nickel content, as shown in Figure 6-12 (33). No explanation has been provided for the maxima in depth of penetration at 0.04% and 0.06% nitrogen.

Cold work is generally considered to decrease the tendency toward sensitization. This is attributed to the provision of carbide nucleation sites within the deformed grains that minimizes the amount of carbide precipitated at grain boundaries during sensitizing heat treatments (43). However, as pointed out by Henthorne (11), cold work could lead to complications relating to the formation of martensitic phases which could be preferentially attacked, acceleration of sigma formation at grain boundaries in molybdenum grades, and increases in end-grain attack. The use of cold work to mitigate susceptibility to intergranular attack should therefore be viewed with caution.

There is some evidence to suggest that increasing grain size increases susceptibility to intergranular attack (44). This has been explained as being due to the fact that in small grained material there is more grain surface area and therefore less chance for a continuous network of

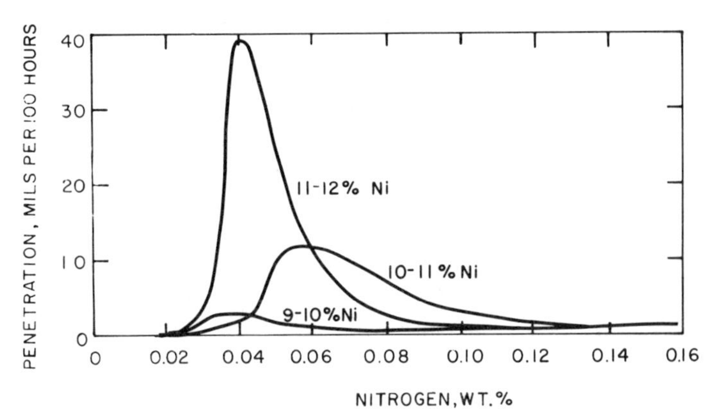

Figure 6-12. Influence of nitrogen content on the depth of intergranular penetration of 18% chromium stainless steels in the Strauss test. Carbon contents in the range 0.02–0.03%. Materials annealed at 1075°C, air cooled, and sensitized at 550°C for 100 hours. (After Binder, Brown, and Franks.)

carbides to form at the grain boundaries (45). It is not clear whether such an effect would be important within the range of grain sizes found in commercial stainless steels.

6-6-2 Ferritic Stainless Steels

The AISI 400 Series. As discussed in Chapter 2, sensitization of the 400 series of ferritic stainless steels is also thought to derive from chromium depletion in the vicinity of precipitated chromium carbides and nitrides at grain boundaries. The temperature range at which sensitization occurs and the kinetics of the process are significantly different from those of austenitic stainless steels, as indicated schematically in Figure 6-13 (13). Sensitization occurs much more rapidly in ferritics.

Types 430, 434, and 446 can be sensitized by heating to temperatures at which there is a significant increase in the solubility of carbon in ferrite. For type 430 this represents temperatures above 925°C (i.e., above line *P-L* in Figure 2-9). Because of the rapid kinetics, sensitization develops in both water quenched and air cooled material (46). The most effective way to eliminate intergranular attack is to anneal these materials at approximately 800°C or to cool very slowly through the temperature range 700–900°C. At these temperatures chromium rediffuses back into the depleted zones. The addition of stabilizing elements (e.g., titanium) minimizes intergranular attack in mild environments, and a titanium stabilized grade, type 439, is commercially available (see Table 2-9). Lowering the carbon content to 0.03% does not prevent attack as it does in austenitic stainless

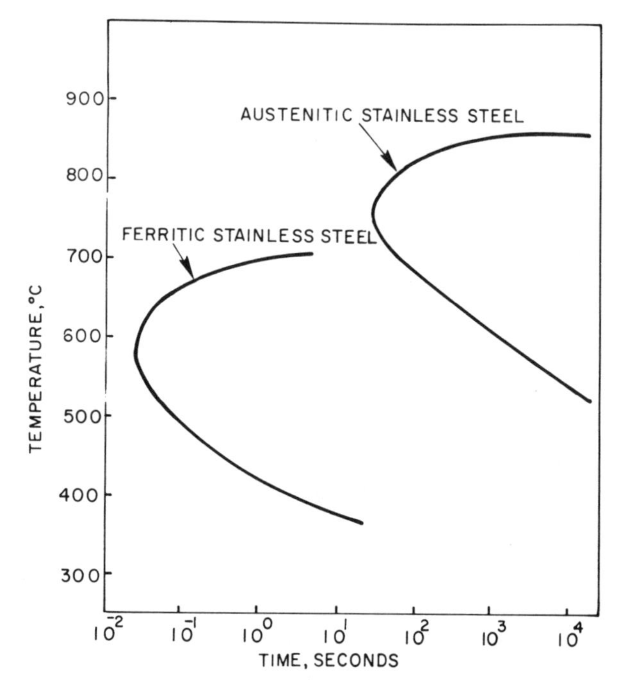

Figure 6-13. Differences in location of time-temperature-sensitization curves of austenitic and ferritic stainless steels of equivalent chromium content. (After Cowan and Tedmon.)

steels. Much lower carbon (plus nitrogen) levels are required to reduce susceptibility to intergranular attack in ferritic stainless steels. The possibility of improving resistance to intergranular attack, as well as improving ductility, by reducing carbon and nitrogen to extremely low levels (e.g., % C + % N = 0.01%) has been one of the stimuli behind the development over the last decade of the new high chromium, low interstitial ferritic stainless steels.

High Chromium Low Interstitial Ferritic Stainless Steels. As noted in Chapter 2, many of these steels are still in the developmental phase, and it is yet uncertain whether the extremely low carbon plus nitrogen contents required for ductility and the extraordinary care required in their welding will inhibit their use as commercial materials of construction for corrosion resistant equipment. However, the studies associated with their development have provided considerable insight into the intergranular corrosion behavior of ferritics and accordingly warrant some discussion. These studies have shown that the key elements that determine the sensitization

TABLE 6-7. RESULTS OBTAINED FROM THE COPPER ACCELERATED STRAUSS TEST ON Fe-17% Cr ALLOYS CONTAINING VARIOUS AMOUNTS OF CARBON AND NITROGEN

Alloy	%C	%N	%N+%C	Temperature, °C of Final Heat Treatment[a]	Bend Test Result[b]
1	0.0021	0.0095	0.0116	788	P
				927	C
				1038	P
				1149	C
2	0.0025	0.022	0.0245	788	P
				927	F
				1038	F
				1149	F
3	0.0031	0.032	0.0351	788	P
				927	F
				1038	F
				1149	F

[a] Water quenched from heat treatment temperature.
[b] P = no cracks visible at 20X after bending.
C = small cracks visible.
F = specimen fractured.

behavior of ferritics are carbon and nitrogen, and that chromium content and cooling rate are also important parameters.

Regarding carbon and nitrogen, Bond (47) has shown that for laboratory prepared high purity Fe-17% Cr alloys, the sum of the carbon plus nitrogen contents has to be of the order of 0.01% to resist sensitization, as given in Table 6-7. The carbon plus nitrogen contents of the higher chromium alloys, such as 26-1, 29-4, and 29-4-2 (see Table 2-11), must also be maintained at very low levels to prevent sensitization. It has been noted that for 26-1 and 29-4, % C + % N should not exceed 0.01% and 0.025%, respectively (48).

However, for the lower chromium materials, even with these low carbon plus nitrogen levels, resistance to sensitization is obtained only if the materials are water quenched from the annealing temperature, as shown in Table 6-8 (49), or reannealed at approximately 788°C and water quenched, as given in Table 6-7. Air cooling can sensitize these materials, whereas the effect of furnace cooling depends on the molybdenum content of the alloys (Table 6-8). The reason for the high corrosion rates observed in furnace cooled molybdenum containing alloys are not fully understood, but are thought to be associated with the effect of molybdenum on nitrogen diffusion rates (49). As demonstrated in Table 6-8, high chromium contents are clearly beneficial in reducing sensitization.

TABLE 6-8. CORROSION RATES IN HUEY TEST AS A FUNCTION OF ALLOY COMPOSITION AND COOLING RATE FROM 1010°C

Alloy Composition, Wt. %	Corrosion Rate, Microns per Year		
	Water Quenched (280°C/S)*	Air Cooled (4.6°C/S)*	Furnace Cooled (0.018°C/S)*
Fe-17.1Cr-0.0020C-0.0060N	439	5010	931
Fe-19.3Cr-0.0020C-0.0040N	323	3520	491
Fe-26.8Cr-0.0010C-0.0081N	138	377	146
Fe-17.3Cr-0.0018C-0.0062N-0.84Mo	545	2760	3570
Fe-19.4Cr-0.0011C-0.0064N-1.01Mo	306	1990	1780
Fe-26.8Cr-0.0012C-0.0092N-1.18Mo	122	337	304

*Cooling rates through the region 600-475°C.

Studies of the effects of stabilizing element additions to Fe-18% Cr alloys containing about 0.06% (C + N), have shown that both columbium and titanium are effective in preventing intergranular corrosion in the copper accelerated Strauss test (50). In the Huey test intergranular attack was observed in the titanium stabilized materials, presumably because of corrosion of titanium carbide by the nitric acid solution. Columbium stabilized material did not exhibit intergranular attack in the Huey test (50). A titanium stabilized grade of 26-1 is commercially available (see Table 2-11).

From a practical viewpoint, it should be emphasized that if resistance to intergranular corrosion is to be maintained for the high purity ferritics, such as 26-1, it is essential that carbon and nitrogen are not picked up during processing and welding. With regard to the latter, scrupulous care must be taken not to contaminate the weld with nitrogen (from air) and carbon (from organics or greases). The welding procedures used for austenitic stainless steels are generally not sufficiently protective for the high purity ferritic stainless steels, and special welding procedures have to be used.

6-6-3 Duplex Stainless Steels

Duplex stainless steels, which contain both austenite and delta ferrite exhibit high resistance to intergranular corrosion in both the Huey test, as given in Table 6-9 (51) and the Strauss test, as illustrated in Figure 6-14 (52). Colombier and Hochmann (52) have suggested a number of reasons why the presence of ferrite in the structure should have a beneficial role. These include the view that the presence of ferrite grains minimizes the formation of continuous networks of chromium carbide along the austenite grain boundaries. Carbides precipitated around isolated ferrite grains

TABLE 6-9. CORROSION RATES IN HUEY TEST OF DUPLEX STAINLESS STEEL IN-744[a]

Condition	Corrosion Rate, inches per month
MA[b]	0.0003
MA + TIG Weld[c]	0.0003
MA + 1 h/482°C/AC	0.0004
MA + 1 h/650°C/AC	0.0004

[a] Fe-26.5%Cr-6.62%Ni-0.06%C-0.20%Ti.
[b] Mill Annealed = cold rolled, annealed at 870°C.
[c] 1.57 mm sheet, autogenous weld.

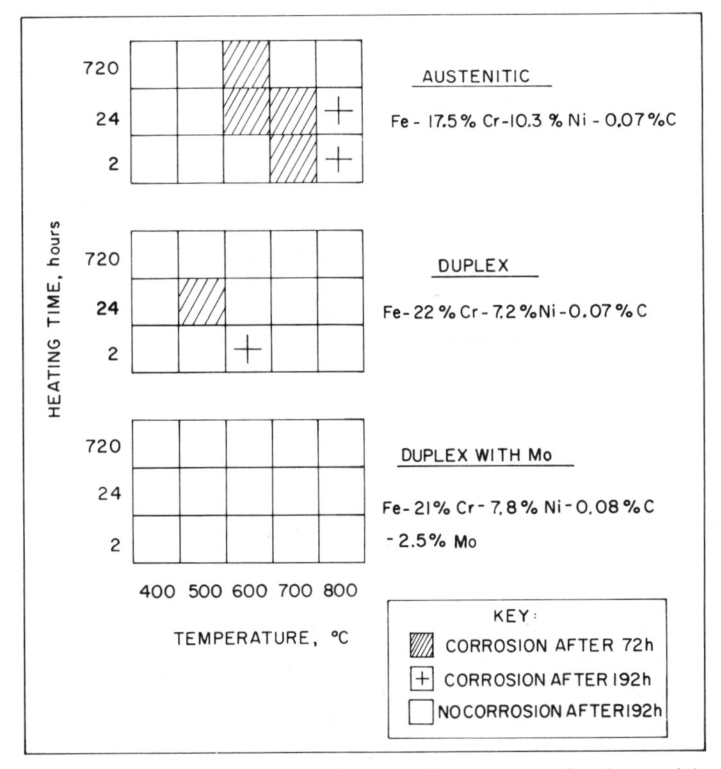

Figure 6-14. Effect of heat treatment on intergranular corrosion of various stainless steels in the Strauss test. Materials water quenched from 1150°C before heat treatment. (After Colombier and Hochmann.)

may cause chromium depletion, but attack will be confined to these isolated grains and will not propagate. Grain boundary migration, observed at 600–800°C (53), may also contribute to the prevention of the formation of continuous paths of chromium depleted material around carbides.

If the intergranular corrosion resistance of duplex stainless steels derives from grain morphology, it would be expected that microstructures containing continuous networks of delta ferrite would exhibit poorer resistance to intergranular corrosion. In this regard Henthorne (11) has reviewed a number of cases where welds of types 316 and 316L, containing continuous networks of delta ferrite, have shown attack in both reducing and oxidizing environments. Very careful consideration should be given to such attack, however, since a number of metallurgical effects could be operative, including molybdenum depletion at delta ferrite/austenite interfaces, chromium depletion, carbide precipitation, and the formation of sigma.

More recent studies* suggest that the high resistance of duplex stainless steels to sensitization stems primarily from a rapid replenishment of chromium due to the more rapid diffusion of chromium in the ferrite phase.

The commercial duplex steels listed in Table 2-5 in the annealed condition (954–982°C/WQ) will pass the Huey test, exhibiting corrosion rates of less than 48 mpy (14). Cowan and Tedmon (13) report that types 329 and 326 may be welded with duplex filler materials with no fear of causing weld decay or knife-line attack. However, mechanical properties could be adversely affected by the formation of alpha prime (475°C embrittlement) and sigma phase in the ferrite. As noted in the subsequent chapter dealing with stress corrosion cracking, the presence of alpha prime also has an adverse effect on stress corrosion resistance of ferritic stainless steels.

The behavior of welds in duplex stainless steels in corrosive environments is a subject that warrants further study.

6-7 INTERGRANULAR CORROSION OF ANNEALED (NONSENSITIZED) STAINLESS STEELS

Annealed austenitic stainless steels are resistant to intergranular attack in almost all environments. However, in certain highly oxidizing nitric acid solutions containing certain metallic ions in their higher valence states, annealed stainless steels can undergo a form of intergranular corrosion which is not linked to sensitization. In terms of the electrochemical

*H. D. Solomon and T. M. Devine, *General Electric Company Report No. 78CRD085,* May 1978.

parameters defined in Chapter 3, this form of attack develops in the transpassive potential range (defined in Figure 3-3) and not in the passive potential range where stainless steels are generally used.

Most studies of the metallurgical aspects of this type of attack have been carried out using corrosion tests in boiling nitric acid solutions containing the hexavalent chromium ion (Cr^{+6}), for example, 5 *N* HNO_3 – 0.5 *N* K_2 Cr_2O_7.

The importance of residual elements in determining the intergranular corrosion resistance of annealed stainless steels was first emphasized by the studies of Chaudron (54), who found that high purity Fe-18% Cr-10% Ni alloys were immune to attack in 11 *N* HNO_3 – 0.57 *N* Cr^{+6} solutions. Subsequent studies by Armijo (55) confirmed Chaudron's findings, and by the use of high purity alloys containing controlled levels of various residual elements, identified phosphorus and silicon as having deleterious effects on intergranular corrosion resistance, as illustrated in Figures 6-15 and 6-16. Increasing the phosphorus content results in a progressive increase in intergranular attack (Figure 6-15). The effect of silicon is more complex. No intergranular attack is observed at low silicon levels (less than 0.1%) or at high silicon levels (greater than 2%), but a maximum rate of attack occurs at approximately 0.7% silicon (Figure 6-16).

While there is some degree of acceptance for the view that the effects of phosphorus and silicon are related to their segregation to grain bound-

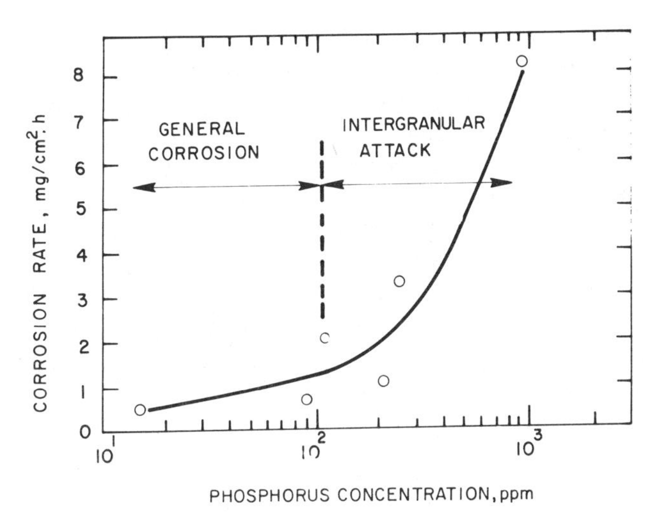

Figure 6-15. Effect of phosphorus content on the corrosion resistance of annealed high purity Fe-14% Cr-14% Ni alloys in nitric acid containing hexavalent chromium. (After Armijo).

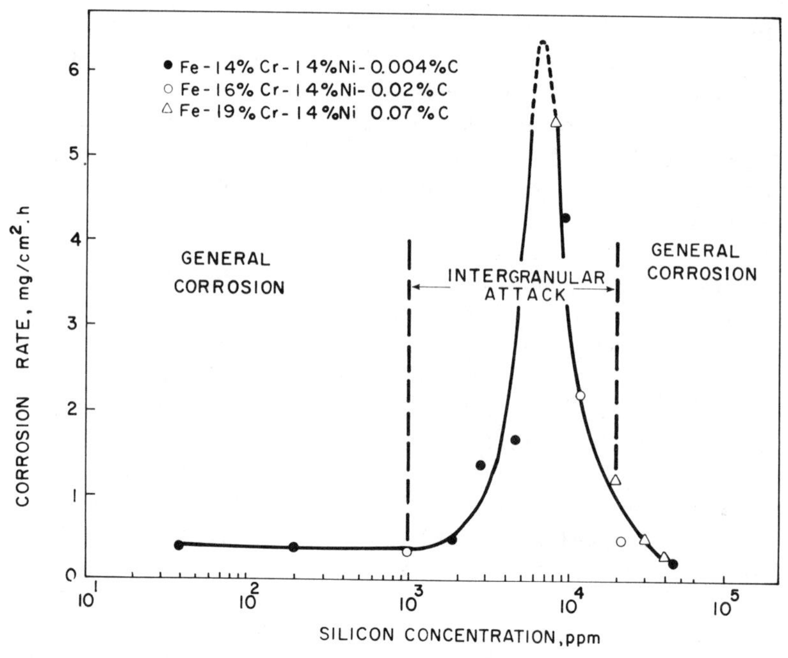

Figure 6-16. Effect of silicon content on the corrosion resistance of annealed high purity Fe-Cr-Ni alloys in nitric acid containing hexavalent chromium. (After Armijo).

aries, the mechanism of attack is not fully understood. For a discussion and critique of the various mechanisms proposed the reader should consult the reviews by Henthorne (11) and Cowan and Tedmon (13).

From a practical viewpoint, the intergranular corrosion of annealed (nonsensitized) stainless steels has not been reported to be a technological problem. The foregoing suggests that high silicon stainless steels should provide resistance to this type of attack in highly oxidizing nitric acid containing high valence metal ions, Cr^{+6} ($Cr_2O_7^{-2}$), Mn^{+7} (MnO_4^{-}), Fe^{+3}, Ce^{+4}, and V^{+5} (56, 57). The low extremes of phosphorus (0.01%) and silicon (0.1%) required for resistance to this type of attack (Figures 6-15 and 6-16) may be difficult to attain with commercial steelmaking practices using stainless steel scrap in the melt charge.

REFERENCES

1. *Intergranular Corrosion of Chromium-Nickel Stainless Steels*, Welding Research Council Bulletin No. 138, February 1969.
2. K. Osozawa and H. J. Engell, *Corros. Sci.*, Vol. 6, p. 389, 1966.
3. W. H. Hatfield (discussion), *J. Iron Steel Inst.*, Vol. 127, p. 380, 1933.
4. B. Strauss, H. Schottky, and J. Hinnuber, *Z. Anorg. Chem.*, Vol. 188, p. 309, 1930.

5. L. R. Scharfstein and C. M. Eisenbrown, "An Evaluation of Accelerated Strauss Testing," *Advances in the Technology of Stainless Steels and Related Alloys*, ASTM STP369, American Society for Testing and Materials, 1965, p. 235.
6. H. J. Rocha (discussion), *Stahl Eisen*, Vol. 75, p. 579, 1955.
7. M. A. Streicher, *ASTM Bulletin* No. 188, p. 35, February 1953.
8. M. A. Streicher, *ASTM Bulletin* No. 229, p. 77, April 1958.
9. M. H. Brown, *Corrosion*, Vol. 30, No. 1, p. 1, 1974.
10. W. R. Huey, *Trans. Am. Soc. Steel Treating*, Vol. 18, p. 1126, 1930.
11. M. Henthorne, "Intergranular Corrosion in Iron and Nickel-Base Alloys," *Localized Corrosion—Cause of Metal Failure*, ASTM STP516, American Society for Testing and Materials, 1972, p. 66.
12. D. Warren, *ASTM Bulletin* No. 230, May 1958, p. 45.
13. R. L. Cowan and C. S. Tedmon, "Intergranular Corrosion of Iron-Nickel-Chromium Alloys," *Advances in Corrosion Science and Technology*, Vol. 3, Plenum Press, New York, 1973, p. 293.
14. E. C. Hoxie, "Some Corrosion Considerations in the Selection of Stainless Steel for Pressure Vessels and Piping," *Pressure Vessels and Piping: Decade of Progress*, Vol. 3, The American Society of Mechanical Engineers, New York, 1977.
15. H. F. Ebling and M. A. Scheil, *Advances in the Technology of Stainless Steels and Related Alloys*, ASTM STP369, American Society for Testing and Materials, 1963, p. 275.
16. K. Osozawa, K. Bohnenkamp, and H. J. Engell, *Corros. Sci.*, Vol. 6, p. 421, 1966.
17. W. L. Clarke, V. M. Romero, and J. C. Danko, *Detection of Sensitization in Stainless Steel Using Electrochemical Techniques*, paper presented at NACE Corrosion/77, Preprint No. 180, 1977.
18. M. Henthorne, *Corrosion*, Vol. 30, p. 39, 1974.
19. H. L. Holzworth, F. H. Beck, and M. G. Fontana, *Corrosion*, Vol. 7, p. 441, 1951.
20. H. T. Shirley and J. E. Truman, *J. Iron Steel Inst.*, Vol. 171, p. 111, 1952.
21. V. Cihal, *Localized Corrosion*, National Association of Corrosion Engineers, Houston, Tex., 1974, p. 502.
22. M. G. Fontana and N. D. Greene, *Corrosion Engineering*, McGraw-Hill Book Company, New York, 1967, p. 66.
23. A. Dravnieks and C. Samans, *Proc. Am. Pet. Inst.*, Vol. 37 (III), p. 100, 1957.
24. C. Samans, *Corrosion*, Vol. 20, No. 8, p. 256t, 1964.
25. R. L. Piehl, *Proc. Am. Pet. Inst.*, Vol. 44 (III), p. 189, 1964.
26. A. C. Hart, *Br. Corros. J.*, Vol. 6, p. 164, 1971.
27. C. Husen and C. H. Samans, *Chem. Eng.*, p. 178, January 27, 1969.
28. G. L. Swales, *Inco-Europe*, private communication.
29. C. D. Stephens and R. C. Scarberry, *Relationship of Polythionic Acid Cracking Susceptibility to Sensitization in Incoloy Alloy 800 and 801*, paper presented at NACE Corrosion/69, Preprint No. 10, 1969.
30. J. J. Heger and J. L. Hamilton, *Corrosion*, Vol. 11, p. 22, 1955.
31. V. Cihal, *Prot. Met.*, Vol. 4, No. 6, p. 563, 1968.
32. A. J. Sedriks, J. W. Schultz, and M. A. Cordovi, *Corrosion Engineering (Boshoku Gijutsu)*, Japan Society of Corrosion Engineering, Vol. 28, p. 82, 1979.
33. W. O. Binder, C. M. Brown, and R. Franks, *ASM Trans. Q.*, Vol. 41, p. 1301, 1949.
34. M. A. Streicher, *J. Electrochem. Soc.*, Vol. 106, p. 161, 1959.
35. D. Warren, *Corrosion*, Vol. 15, p. 213t, 1959.
36. V. Cihal, I. Kasova, and J. Kubelka, *Metaux Corros. Ind.*, No. 529 (September), p. 281, 1969.

37. P. Schwaab, W. Schwenk, and H. Ternes, *Werkst. Korros.*, Vol. 16, p. 844, 1965.
38. A. C. Hamstead and L. S. Van DeLinder, *Corrosion*, Vol. 15, p. 147t, 1959.
39. C. P. Dillon, *Corrosion*, Vol. 16, p. 433t, 1960.
40. R. Franks, *Corrosion Handbook*, John Wiley, New York, p. 160, 1948.
41. M. H. Brown, W. B. DeLong, and W. R. Meyers, *Evaluation Tests for Stainless Steels*, ASTM STP93, American Society for Testing and Materials, 1950, p. 103.
42. G. Lennartz, *Mikrochim. Acta*, Vol. 3, p. 405, 1965.
43. E. C. Bain, R. H. Aborn, and J. B. Rutheford, *Trans. Am. Soc. Steel Treat.*, Vol. 21, p. 481, 1933.
44. H. D. Newell, *Trans. Am. Soc. Steel Treat.*, Vol. 19, p. 673, 1932.
45. V. Cihal, *Protection Mater.*, Vol. 2, No. 2, p. 127, 1966.
46. R. A. Lula, A. J. Lena, and G. C. Kiefer, *Trans. ASM*, Vol. 46, p. 197, 1954.
47. A. P. Bond, *Trans. AIME*, Vol. 245, p. 2127, 1969.
48. *Materials Engineering*, Manual 267, *Ferritic Stainless Steels*, April 1977, p. 69.
49. R. J. Hodges, *Corrosion*, Vol. 27, p. 119, 1971.
50. A. P. Bond and E. A. Lizlovs, *J. Electrochem. Soc.*, Vol. 116, p. 1305, 1969.
51. *Preliminary Data: IN-744 Stainless Steel*, The International Nickel Company, October 1, 1969.
52. L. Colombier and J. Hochmann, *Stainless and Heat Resisting Steels*, St. Martin's Press, New York, 1968, p. 162.
53. C. S. Tedmon and D. A. Vermilyea, *Met. Trans.*, Vol. 1, p. 2045, 1970.
54. G. Chaudron, *EURAEC-976 Quarterly Report No. 6*, October/December 1963.
55. J. S. Armijo, *Corrosion*, Vol. 24, p. 24, 1968.
56. M. A. Streicher, *J. Electrochem. Soc.*, Vol. 106, p. 161, 1959.
57. H. Coriou, J. Hure, and G. Plante, *Electrochim. Acta*, Vol. 5, p. 105, 1961.

7

STRESS CORROSION CRACKING

7-1 INTRODUCTION

Stress corrosion cracking (SCC) is a general term describing stressed alloy failures that occur by the propagation of cracks in corrosive environments. SCC has the appearance of brittle fracture, yet it can occur in highly ductile materials. The requirements for SCC to occur are the presence of a tensile stress, either residual, applied, or a combination of both, and the presence of a specific corrodent. The cracks form and propagate roughly at right angles to the direction of the tensile stress at stress levels much lower than those required to fracture the material in the absence of the corrodent. On a microscopic scale the cracks that run across grains are called transgranular and those that follow grain boundaries are termed intergranular, as illustrated in Figure 7-1. When SCC has progressed to a depth at which the remaining load bearing section of a material reaches its fracture stress in air, the material separates by normal overload fracture—generally by microvoid coalescence in ductile materials. Thus the fracture surface of a component failed by SCC will contain areas characteristic of SCC as well as areas exhibiting the "dimples" associated with microvoid coalescence.

Much of the literature on the subject of SCC of stainless steels has dealt with attempts to understand the mechanism of failure. The suggested mechanisms can be categorized according to the unit process thought to be responsible for the propagation of the stress corrosion crack. There are two basically different unit processes: (*a*) the removal of material at the crack tip, and (*b*) the separation of material at the crack tip. These unit processes provide the basis for the classification of the SCC mechanisms shown in Table 7-1 (1). The so-called martensite mechanisms, which can be fitted either into the dissolution or hydrogen embrittlement categories, rely on the formation of martensite, during straining, in the vicinity of the

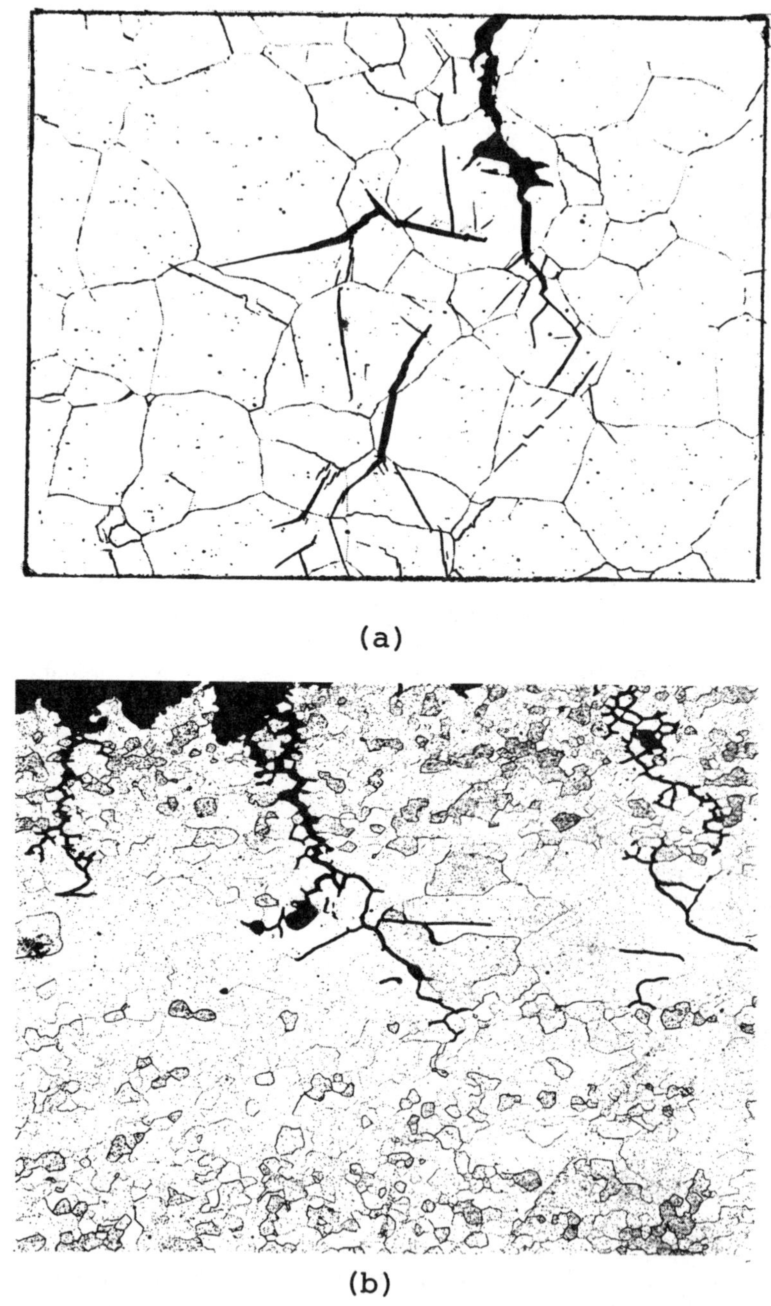

Figure 7-1. (*a*) Transgranular cracks in an austenitic stainless steel produced in a chloride environment, 100×; (*b*) intergranular cracks in a ferritic stainless steel produced in a high temperature caustic environment. 50×.

crack tip. As pointed out by Staehle (2), strain induced martensite cannot explain SCC at elevated temperatures, since SCC of type 304 can occur at temperatures much higher than the M_d temperature for type 304. However, the presence of strain induced martensite in type 301 can lead to hydrogen embrittlement at ambient temperatures when this material is galvanically coupled to corroding zinc or magnesium anodes (3).

The multiplicity of mechanisms proposed (Table 7-1) and the lack of

TABLE 7-1. SUGGESTED CLASSIFICATION OF STRESS CORROSION CRACKING MECHANISMS

DISSOLUTION MECHANISMS

Film Rupture

Crack propagates by local dissolution of metal at crack tip due to prevention of passivation there by plastic deformation.

Stress Accelerated Dissolution

Crack propagates by localized anodic dissolution. Principal role of plastic deformation is to accelerate the dissolution process.

MECHANICAL MECHANISMS

Hydrogen Embrittlement

Hydrogen accumulates within metal in the crack-tip region, leading to localized weakening either by void formation or lowering of cohesive strength. Crack propagates by mechanical fracture of weakened region.

Adsorption

Surface-active species adsorb and interact with strained bonds at the crack tip, causing reduction in bond strength and leading to crack propagation.

MIXED MECHANISMS

Brittle Film

Crack propagates by repeated formation and rupture of a brittle film which grows into metal at crack tip.

Tunnel Model

Crack propagates by formation of deep pits or tunnels via dissolution followed by linking of these pits or tunnels by ductile rupture.

agreement as to which mechanisms are operative in stainless steels has led to the evolution of a simplistic nomenclature to describe SCC in terms of the environments that cause it. Thus the terms chloride cracking, caustic cracking, and oxygen cracking are often used to describe SCC of stainless steels in elevated temperature water environments containing chloride, caustic, or oxygen. The term hydrogen embrittlement is used to describe cracking in environments in which the cracking process is facilitated by the entry of hydrogen into the metal. The term sulfide cracking usually denotes hydrogen embrittlement, with the sulfide ion acting as a hydrogen ion recombination poison promoting the entry of hydrogen into the metal. To what extent these variously described forms of cracking are manifestations of the same unit process of crack propagation remains a matter of continuing discussion which is not likely to be settled in the near future. Accordingly, in this chapter the various forms of cracking are discussed under the environment related descriptions noted above.

7-2 CHLORIDE CRACKING

7-2-1 General Background

The fact that stressed stainless steels can exhibit SCC in elevated temperature chloride solutions has been known for some 40 years (4), and the topic has received much laboratory study and publicity. However, it should be remembered that some 80% of the stainless steel produced is used by industries which do not report a significant number of SCC failures (Figure 1-3). Since chloride cracking in stainless steels was first noted and subsequently widely studied in the austenitic grades, a misleading impression has sometimes emerged that chloride cracking is associated only with austenitic stainless steels and that ferritic and duplex stainless steels are immune to this form of cracking. As shown in the subsequent discussion, there is no immunity associated with the body centered cubic ferrite lattice, and, as in the case of austenitics, susceptibility to chloride cracking is dependent on composition and structure.

While other forms of localized attack in chloride solutions are associated with structural heterogeneities (e.g., pitting often initiates at manganese sulfide inclusions and intergranular corrosion at chromium depleted grain boundaries), chloride cracking usually adopts a transgranular path, with some segments of the crack following the slip planes of the austenite lattice (5, 6). Accordingly, there have been attempts to characterize chloride cracking in terms of deformation behavior, specifically by considering events that are likely to occur when the slip intersects the surface at the crack tip. This creates a slip step that can rupture the passive film and expose a fresh metal surface to the corrosive environment. These theories are described under the heading of "film rupture" in Table 7-1.

The concept behind the film rupture theory is generally attributed to Champion (7) and essentially entails a competition between the tendency of the material to passivate by the formation of a protective film, and the prevention of the development of such a film by the formation of slip steps which rupture it. As pointed out by Parkins (8), this model can give rise to a transgranular crack path. Thus, even if the initial crack nucleus is formed at a grain boundary site, the rupture of the film by slip steps would be expected to redirect the crack path across grains. However, Sedriks et al. (9) have noted that intergranular cracking would be expected in a situation in which grain boundaries are less readily passivated than the grains. Thus, in a growing crack initiated at a grain boundary, ruptured film sites further away from the boundary would passivate more readily than ones near the boundary. The net effect would be to maintain crack propagation near or at the grain boundary. This could explain the observation that chloride cracking in heavily sensitized stainless steels, in which the chromium depleted grain boundaries are less readily passivated than the grains, is often intergranular (10).

While the film rupture model cannot as yet be quantified, it appears to be a likely model for explaining chloride cracking in stainless steels, since it can at least qualitatively explain the effect of some parameters observed with this form of cracking. For example, it predicts that increasing the stress level will increase susceptibility to cracking, and that increasing the content of alloying elements that enhance passivity, such as nickel, chromium, and molybdenum, to high levels will result in increased cracking resistance or immunity. Because the passivating tendency must be sufficiently strong to instantly heal mechanical ruptures of the passive film in the acidic environment at the crack tip, these alloying element levels must be quite high. For example, recent research at Inco (11) suggests that in the Fe-Cr-Ni alloy system, even at the 30% chromium level, the nickel content must be raised to 60% to ensure against cracking in high temperature water environments, as well as in boiling concentrated chlorides. The outcome of these studies has been the development of a commercial alloy designated Inconel alloy 690 (Table 2-7), which has not exhibited chloride cracking in any heat treated or welded condition tested to date (11).

However, because of the many complex factors associated with chloride cracking, the generally held view is that it is an oversimplification to regard this phenomenon only in terms of slip and repassivation. While it is true that alloys containing high amounts of nickel, chromium, and molybdenum are resistant to chloride cracking, the beneficial effect of nickel on cracking resistance is far greater than would be expected from its effect on pitting potential (Figure 4-7) whereas molybdenum, which has a very great effect on pitting potential (Figure 4-8), can in fact be detrimental to cracking resistance at lower alloying levels. The view has emerged, therefore, that metallurgical factors that affect slip processes, mechanical factors, such as strain rate and stress intensity, film parameters, such as thickness, strength, and plasticity, and a multitude of environmental parameters must be considered in attempts to construct a quantifiable (predictive) model for chloride cracking. For an assessment of the state-of-the-art in mechanistic studies the reader should consult references 12 and 13.

7-2-2 Austenitic Stainless Steels

The Magnesium Chloride Test. Because chloride cracking was first noted in the more widely used austenitic stainless steels, these materials have received by far the greatest amount of study. In the majority of cases the evaluations have employed boiling magnesium chloride solutions. These are very severe environments. Nevertheless, like the Huey test for intergranular corrosion, the magnesium chloride test has remained popular because of its severity, and past and present attempts to introduce less severe tests such as the Wick test and the boiling 26% sodium chloride test have not been met with a great deal of acceptance. The popularity of the boiling magnesium chloride test derives from the belief among end users that stainless alloys that survive lengthy periods of exposure to concentrated boiling magnesium chloride are not likely to exhibit chloride cracking in service. It has also been favored by researchers because it is simple and rapid, and provides a comparison among tests in different laboratories.

The boiling magnesium chloride test was first described by Scheil in 1945 (14). It has now been standardized as ASTM G36-73, based on studies by Streicher and Sweet (15) and Casale (16). The boiling point of magnesium chloride solutions is strongly dependent on concentration, as shown in Figure 7-2 (16), and ASTM G36-73 recommends the use of a test solution that boils at 155°C $\pm$1°C. It is important to make alloy comparisons under the same conditions, since, as shown by Kowaka and Kudo (17), significant differences in the relative susceptibilities of type 304 and 316 are found at different temperatures or concentrations, as illustrated in Figure 7-3 (17). ASTM G36-73 notes that any type of stress corrosion test specimen can be used. For a comprehensive discussion of the various types of test specimens the reader may consult reference 18. A popular test specimen has been the U-bend specimen described in ASTM G30-72. The U-bend specimen contains large elastic and plastic strains and provides one of the most severe configurations available with smooth (as opposed to notched or fatigue precracked) test specimens. It is also

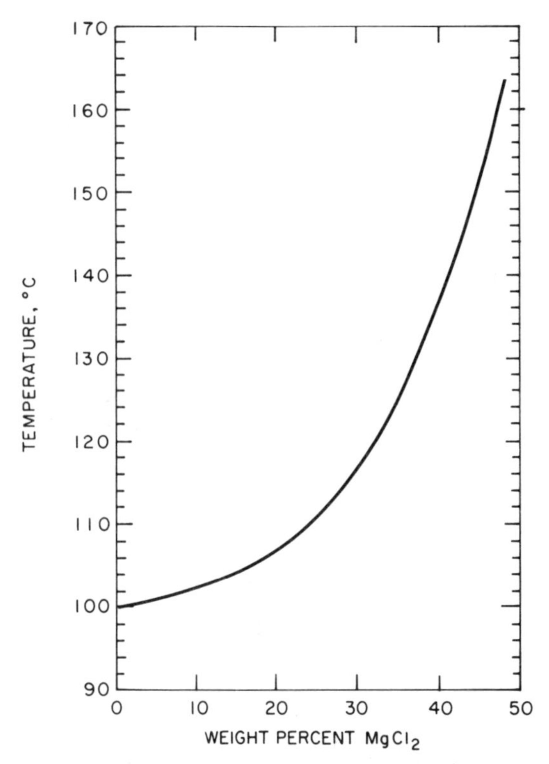

Figure 7-2. Boiling points of aqueous magnesium chloride solutions at one atmosphere as a function of concentration. (After Casale.)

simple and relatively inexpensive to make and use, as demonstrated in Figure 7-4.

Effects of Composition. As noted before, there has been an extensive research effort, often employing boiling magnesium chloride environments, to investigate the effects of alloying additions on the chloride cracking resistance of stainless steels. The results, summarized in Figure 7-5, have been taken largely from reviews by Latanision and Staehle (19) and Theus and Staehle (13). It is evident from Figure 7-5 that many alloying additions appear to be detrimental to chloride cracking, but there are also those which can be categorized as "beneficial" and "variable." Those designated beneficial in Figure 7-5 are nickel, cadmium, zinc, silicon, beryllium, and copper. It should be emphasized that the beneficial effect of nickel relates to austenitic stainless steels. When present in relatively small quantities in ferritic stainless steels, nickel is detrimental to chloride cracking resistance. The beneficial effect of nickel on chloride cracking resistance of austenitic stainless steels and higher alloys has

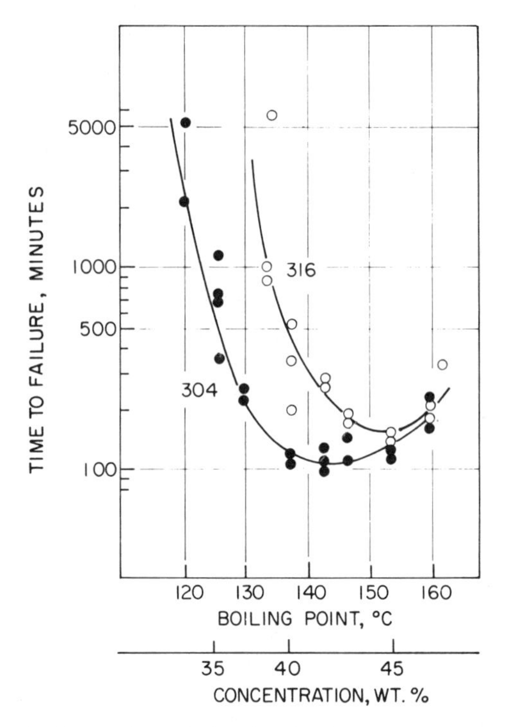

Figure 7-3. Times to failure of type 304 and 316 stainless steels at a stress of 245 MPa in various boiling magnesium chloride solutions. (After Kowaka and Kudo.)

been extensively studied and is well documented. Useful examples are the studies by Copson (20), Staehle (21), and Denhard and Gaugh (22), the results of which are illustrated in Figures 7-6–7-8, respectively. These evaluations were carried out in boiling magnesium chloride solutions. More recent studies reported by Sedriks (23), summarized in Table 7-2, have shown that the beneficial effect of nickel is by no means confined to magnesium chloride environments, but is in evidence in other chloride environments as well.

Additions of cadmium and zinc are also designated as beneficial (Figure 7-5). Actually there are very little data for these alloying elements. It has been reported (24) that a 0.2% cadmium addition to type 304 inhibits cracking in a vapor test which readily cracked type 304. Regarding zinc, Royuela and Staehle (25) have shown an improvement by a factor of five resulting from a 1% zinc addition to an Fe-20% Cr-15% Ni alloy. No systematic studies have been reported for cadmium and zinc additions, and hence the designation of these elements as beneficial must be regarded as tentative.

The evidence for the beneficial effect of silicon and beryllium is some-

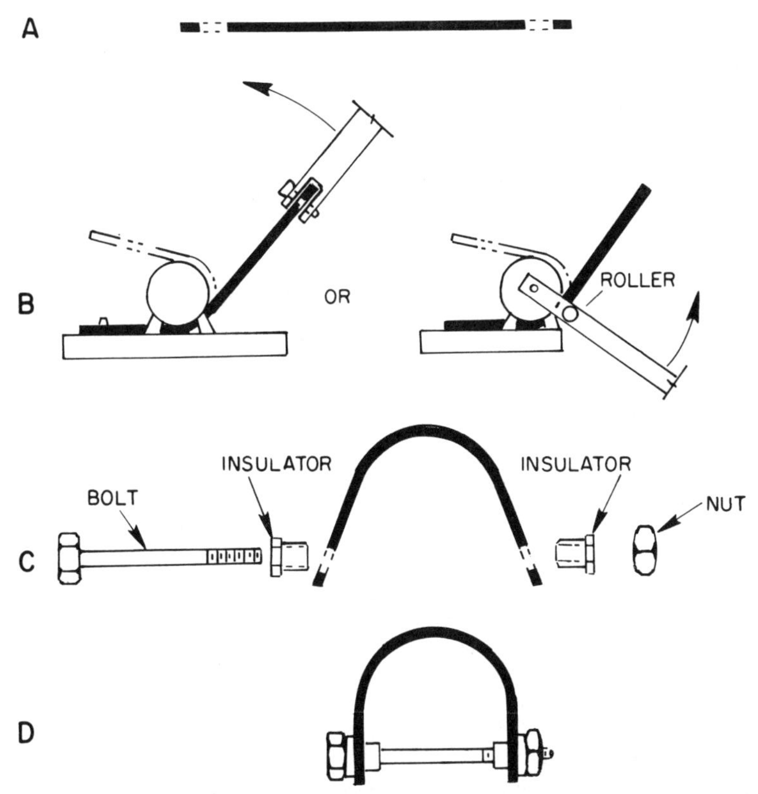

Figure 7-4. Preparation of U-bend stress corrosion test specimen: *A*, specimen blank; *B*, bending; *C*, components for assembly; *D*, assembled specimen. (ASTM G30-72.)

what more extensive. The effect of silicon has been studied by a number of investigators, with the results of Bourrat and Hochman (27) given in Table 7-3. There appears to be no controversy regarding the beneficial effect of silicon on chloride cracking resistance in boiling 42% magnesium chloride. However, studies by Truman (28), summarized in Table 7-4, show that the beneficial effect of silicon noted in magnesium chloride tests is barely detectable, if at all, in sodium chloride solutions at higher temperatures.

Regarding beryllium, Staehle et al. (29) have shown that the additions of this element to an Fe-20% Cr-15% Ni alloy are beneficial. The beneficial effect of copper is slight (29).

The elements designated in Figure 7-5 as having a variable effect on chloride cracking resistance can be divided into three categories as shown in Table 7-5. Boron, aluminum, and cobalt appear to be detrimental in small quantities, but beneficial in larger quantities (25, 26, 29, 30–32); tin

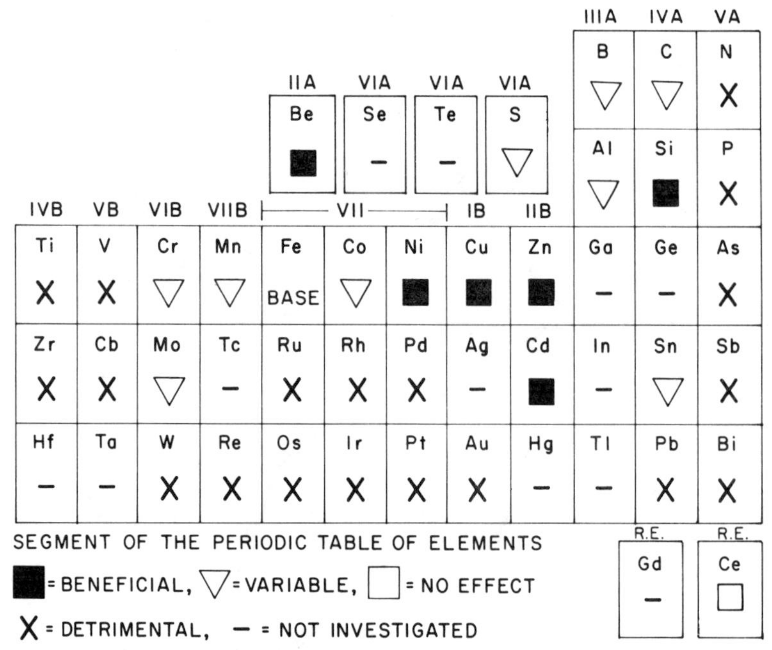

Figure 7-5. Effect of element shown on resistance of austenitic stainless steels to stress corrosion cracking in chloride solutions.

and manganese appear to have no effect in certain ranges, and beneficial and detrimental effects in other ranges (19, 32); carbon, molybdenum, and chromium show minima in chloride cracking resistance. Hines and Jones (33) have suggested that the carbon minimum lies somewhere in the range 0.06–0.10%, while the molybdenum minimum is near 1.5% with the exact value depending on the carbon content. Latanision and Staehle (19) have attempted to analyze the published data showing the effect of chromium variation and have concluded that for alloys containing 10–20% nickel, the chromium minimum lies somewhere between 12 and 25%, probably near 20%.

Examination of the literature reveals that of the three alloying elements that enhance resistance to localized attack in austenitic materials, namely, nickel, chromium, and molybdenum, only the beneficial effect of nickel is well defined for chloride cracking. A better definition, at fixed nickel contents, of the beneficial ranges of chromium and molybdenum appears to be desirable. Also, as noted in Chapter 5, the role of the formation of the molybdate stabilized salt film should be investigated.

The foregoing survey of the effects of alloying additions on the chloride cracking resistance of austenitic stainless steels pertains to materials made by conventional melting practice, which contain many impurities.

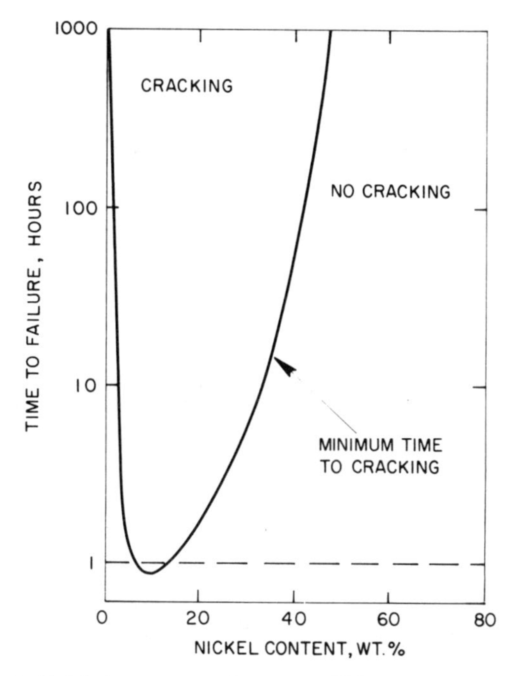

Figure 7-6. Effect of nickel content on the susceptibility to stress corrosion cracking of stainless steel wires containing 18–20% chromium in a magnesium chloride solution boiling at 154°C. (After Copson.)

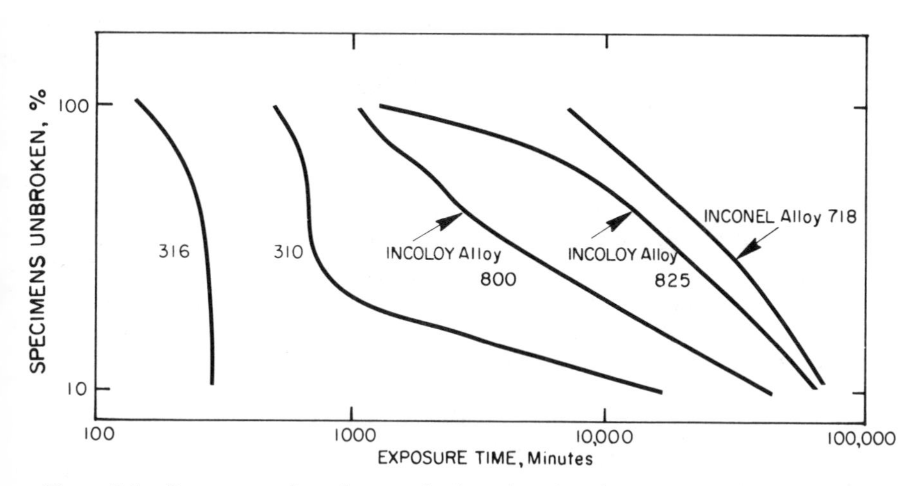

Figure 7-7. Percentage of specimens unbroken plotted against exposure time for various commercial alloys exposed to a magnesium chloride solution boiling at 154°C. Specimens stressed at 90% of the yield stress. (After Staehle.)

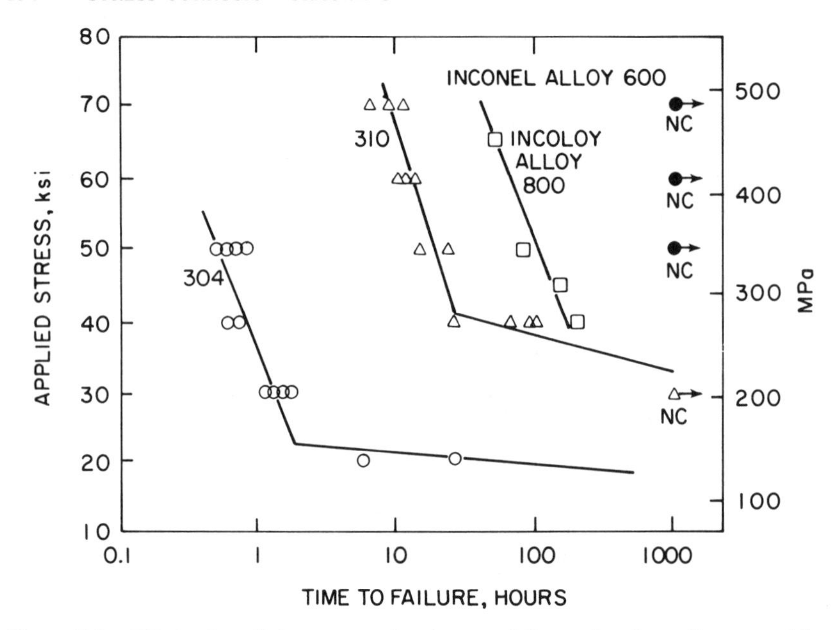

Figure 7-8. Effect of applied stress on the times to failure of various alloys tested in a magnesium chloride solution boiling at 154°C. NC denotes no cracking. (After Denhard and Gaugh.)

In this regard it has been shown that austenitic (and, as shown later, also ferritic) stainless steels of very high purity are highly resistant to cracking in boiling magnesium chloride. This has been shown in the case of Fe-16% Cr-20% Ni (32) and Fe-18% Cr-14% Ni (34) high purity alloys. In the latter case, the total metallic impurities were quoted at 1 ppm and the total nonmetal impurities (oxygen, nitrogen, sulfur, and phosphorus) at 10 ppm, the material having been prepared by plasma furnace melting. However, at present the approach of increasing chloride cracking resistance by making very high purity alloys remains a matter of fundamental rather than commercial interest. Nevertheless, such studies have the potential of making a major contribution to the understanding of the role of alloy impurities in chloride cracking which, as indicated in Figure 7-5, can have varied effects on cracking resistance.

Effects of Stress. As shown in Figure 7-8, decreasing the applied stress increases time to failure, and at low stresses a run-out (threshold) stress is indicated. Very careful studies by Spahn et al. (35) using solution annealed type 347 electropolished tensile specimens have established a threshold stress for this material at approximately 160 MPa (23 ksi), as

TABLE 7-2. RESULTS OF STRESS CORROSION TESTS OF Fe-Cr-Ni ALLOYS IN VARIOUS ENVIRONMENTS

Environment	Test Time, days	Test Temp., °C	Average Times to Failure, Days			
			Type 304	Type 310	INCOLOY Alloy 800	INCONEL Alloy 600
3.5% NaCl + 0.5% $CH_3 \cdot COOH$ + H_2S	30	30	5	NF	NF	NF
50% H_2SO_4 + 3% NaCl	30	30	1	NF	NF	NF
NaCl, Wick Test	30	100	3	16	--	NF
Saturated NaCl, pH = 4	30	109	12	NF	NF	NF
36% $CaCl_2$, Boiling	30	110	5	--	NF	--
45% $MgCl_2$, Boiling	10	154	1	1	4.5	NF
85% $ZnCl_2$, Boiling	10	180	1	1	3.5	NF
Aerated Water + 875 ppm NaCl*	56	260	7	--	56	NF

NF = no failure at end of test.
\- = not tested.
* = Autoclave tests, weekly inspection of specimens.

TABLE 7-3. EFFECT OF SILICON ON STRESS CORROSION RESISTANCE IN BOILING 42% MAGNESIUM CHLORIDE

Material	Time to Failure in Boiling 42% $MgCl_2$, hours
Fe - 18%Cr - 14%Ni - 0.06%Si	158
- 1.0%Si	700
- 2.2%Si	193-672
- 3.3%Si	>1000
- 4.0%Si	>1000

indicated in Figure 7-9. The presence of a threshold stress intensity, using precracked fracture mechanics specimens in boiling magnesium chloride, has also been noted (36, 37). For type 304L, this stress intensity (K_{Iscc}) is about 8 MPa$\sqrt{m}$, as shown in Figure 7-10 (37).

A certain amount of consideration has been given to increasing resistance to chloride cracking by reducing stress level. Stress, however, is a difficult parameter to control, and while it is obviously good practice to minimize applied (design) stresses, the presence of tensile residual stresses in the material can negate any benefit. The only sure way to control stresses is by stress relief annealing the fully assembled structure. Full stress relief can be attained by annealing in the range 800–900°C, since this causes a recrystallization of the deformed (internally stressed) grains. Unfortunately, this is often only an option for small components. For larger components it is sometimes possible to achieve partial stress relieving by heating to 400–600°C, and it is believed, particularly in Europe, that even such partial stress relieving can significantly improve resistance to chloride cracking. Any such procedures should be carefully weighed against possible problems associated with sensitization (see Chapter 6).

Compressive residual stresses, for example, such as may be introduced by careful shot peening, have been shown to increase resistance to chloride cracking, as indicated in Figure 7-11 (38).

Effects of Environment. Temperature is by far the most important environmental variable in determining whether chloride cracking will occur. As noted by Truman (39), decades of practical experience have shown that for austenitic stainless steels chloride cracking is not a hazard at ambient temperatures. Recent work by Money and Kirk (40), summarized in Table 7-6, suggests that this applies also to welded austenitic stainless steels in ambient temperature marine atmosphere. As shown in this table, attack can occur in severely furnace sensitized materials, but not in welded materials, except type 301 which often contains martensite. Unlike chloride cracking which is usually transgranular, the attack ob-

TABLE 7-4. EFFECT OF SILICON ON STRESS CORROSION RESISTANCE OF AUSTENITIC STAINLESS STEELS IN VARIOUS CHLORIDE MEDIA

Analysis, Wt. %*						Ferrite Content	Time to Failure, hours		
C	Si	Mn	Cr	Ni	N	%	42% $MgCl_2$ at Boiling Point	100 ppm NaCl at 250°C	1000 ppm NaCl at 250°C
0.020	0.46	0.83	18.28	15.10	0.07	0.5	219	28	10
0.022	0.80	0.77	18.54	15.20	0.08	0.5	292	24	5
0.031	2.48	0.83	18.58	15.23	0.07	1	1000 NF**	92	32
0.031	3.69	0.88	18.46	15.20	0.07	1	1000 NF	81	15
0.031	4.53	0.86	18.42	15.28	0.08	2.5	1000 NF	54	50

*Balance iron.
**1000 NF = no failure in 1000 hours.

TABLE 7-5. ELEMENTS SHOWING VARIABLE EFFECT ON RESISTANCE OF AUSTENITIC STAINLESS STEELS TO CHLORIDE CRACKING

Element	Beneficial	Detrimental
B	>0.1%	0.01%
Al	0.1%	0.04%
Co	>1.8%	1.50%
Sn	0.001-0.02	No Effect at 0.4%
Mn	No Effect 0-2%	>2%
	Minimum in Cracking Resistance at:	
C	0.06-0.1%	
Mo	∿1.5%	
Cr	12-25%	

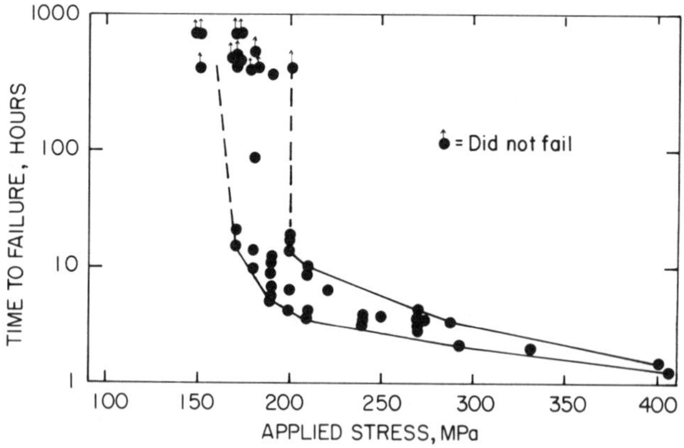

Figure 7-9. Effect of applied stress on the time to failure of solution annealed and electropolished type 347 stainless steel, tested in a magnesium chloride solution boiling at 145°C. (After Spähn, Wagner, and Steinhoff.)

served by Money and Kirk was intergranular and clearly related to sensitization. It seems likely, therefore, that this attack is a form of intergranular corrosion assisted by stress.

Chloride cracking becomes noticeable at elevated temperatures. The traditional engineering viewpoint, based on practical experience (41), is that chloride cracking can occur at temperatures above 60°C (140°F), provided that the material is exposed for very long times. The extreme dependence of chloride cracking on temperature is shown in Figure 7-12 (39).

As indicated in Figure 7-13 (42), the relationship between chloride concentration and cracking susceptibility is not simple, and in high tem-

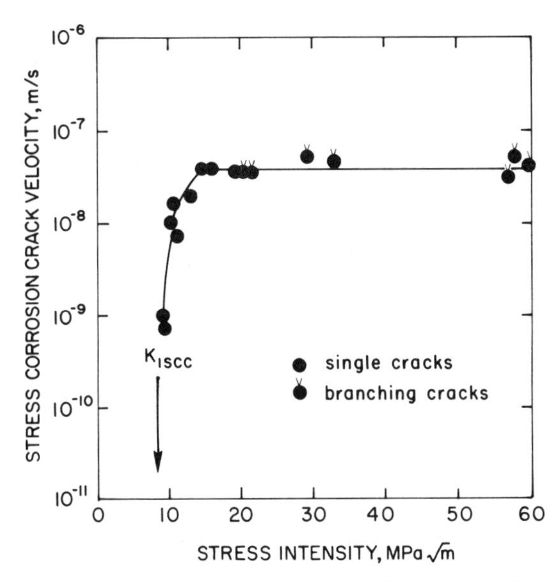

Figure 7-10. Effect of stress intensity on the velocity of stress corrosion cracks in solution annealed type 304L stainless steel tested in a magnesium chloride solution boiling at 130°C. (After Speidel.)

perature environments it is linked with oxygen concentration. It is less clear whether the presence of oxygen is necessary to cause cracking in concentrated chloride solutions boiling at atmospheric pressure, since their oxygen contents should approach zero. However, Figure 7-14 (43) shows that the introduction of oxygen into such solutions by bubbling accelerates cracking. There is some evidence to suggest that in oxygen containing higher temperature sodium chloride solutions, susceptibility to cracking decreases with decreasing chloride concentration, as indicated in Figure 7-15 (39). However, the maintenance of low chloride levels cannot ensure freedom from cracking under conditions in which chloride can concentrate in crevices or shielded areas. The effect of temperature on chloride cracking susceptibility of type 304 under concentrating conditions is shown in Figure 7-16 (44). In the Wick test used to obtain the data of Figure 7-16, a porous insulating material was used as a wick to draw the salt solution to the heated surface of a stressed specimen at which chloride can concentrate. This test was originated by Dana and DeLong (45).

Regarding pH, Baker et al. (46) have determined that the pH of the solution at the tip of a crack in type 304 undergoing cracking in magnesium chloride is between 1.2 and 2.0. This suggests that, as in the case

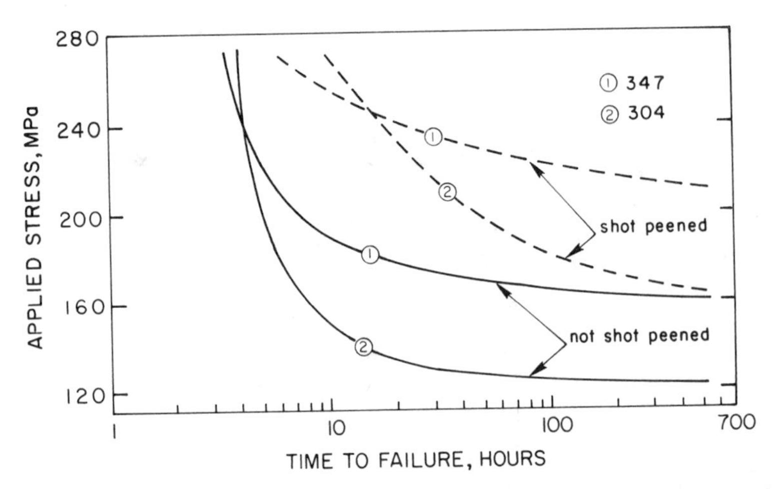

Figure 7-11. Effect of shot peening with 40–80 μm glass shot on the times to failure of type 304 and 347 stainless steels in a boiling 42% magnesium chloride solution. (After Fässler.)

TABLE 7-6. RESULTS OF STRESS CORROSION TESTS IN WHICH U-BEND SPECIMENS WERE EXPOSED FOR FIVE YEARS TO MARINE ATMOSPHERE IN THE 80 FOOT LOT AT KURE BEACH

	Condition			
Material	Annealed	Welded	Cold Worked, 1/4 Hard	Sensitized, 650°C/1.5 h/FC
201	NF	NF	NF	IGA
301	NF	IGA	NF	IGA
302	NF	NF	NF	IGA
304	NF	NF	NF	IGA
304L	NF	NF	NF	NF
309	NF	NF	--	IGA
310	NF	NF	--	NF
316	NF	NF	NF	IGA
CARPENTER 20Cb3	NF	NF	--	NF
INCOLOY alloy 825	NF	NF	--	NF
INCOLOY alloy 800	NF	NF	--	NF
INCONEL alloy 600	NF	NF	--	NF

NF = no failure in five years.
IGA = intergranular attack, possibly stress assisted.
-- = not tested.

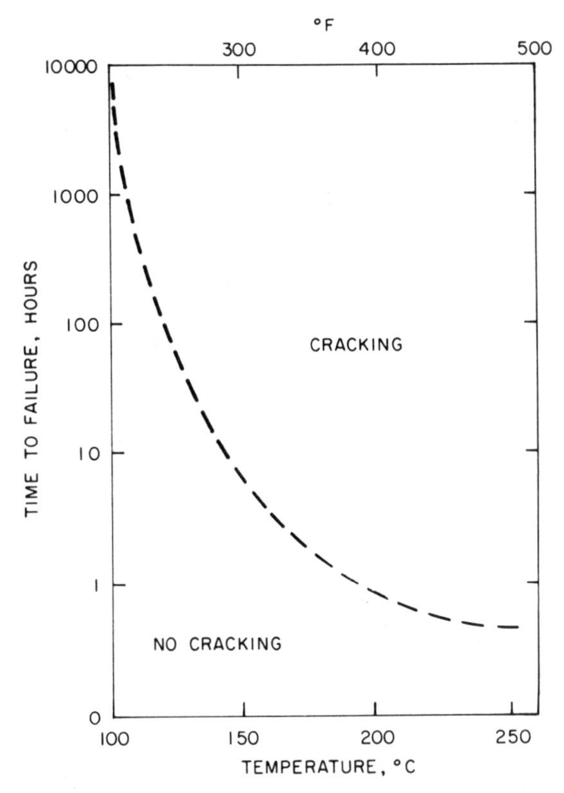

Figure 7-12. Effect of temperature on the chloride cracking resistance of type 304 and 347 stainless steels in 0.1–33.0% sodium chloride containing oxygen. (After Truman.)

of pitting and crevice corrosion, acidification may be occurring by the hydrolysis of metal ions. In the case of pitting, increase in the bulk pH of the solution increases pitting resistance (Figure 4-16), and a similar improvement with increasing pH has been reported for chloride cracking resistance [e.g., Figure 7-17 (47) and Table 7-7 (48)]. More detailed studies of the beneficial effect of increasing alkalinity seem desirable. However, as noted later, in stronger caustic solutions caustic cracking may occur, particularly in sensitized materials.

Among environmental, as opposed to metallurgical, remedies that can be used to minimize or prevent chloride cracking, cathodic protection and the addition of inhibitors deserve mention. Regarding cathodic protection, Couper (49) has shown that both impressed currents and sacrificial anodes can be used to prevent cracking of type 304 in boiling 42% magnesium chloride. More detailed studies by Brauns and Ternes (50), summarized in

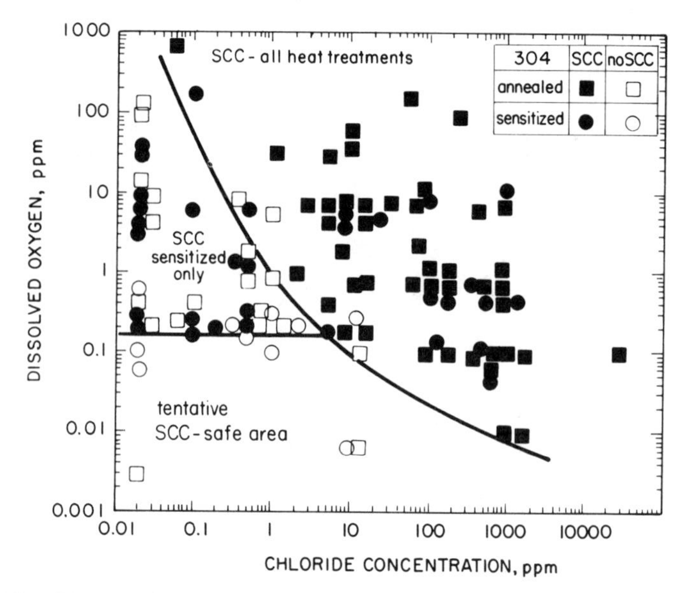

Figure 7-13. Concentration ranges of dissolved oxygen and chloride that may lead to stress corrosion cracking of type 304 stainless steel in water at temperatures in the range 260–300°C. Applied stresses in excess of yield strength and test times in excess of 1000 hours, or strain rates greater than 10^{-5}/second. (After Speidel.)

Figure 7-18, have established the relationships between applied potential, applied stress, and time to failure for type 304 stainless steel in boiling 42% magnesium chloride.

Studies of inhibitors have been more extensive (49, 51–57). It has been found that silicates, nitrates, phosphates, carbonates, iodides, and sulfites are all effective inhibitors of cracking if present in certain concentrations.

Effects of Microstructure. The presence of delta ferrite in austenitic stainless steels generally improves resistance to chloride cracking. An example of this for some cast austenitic stainless steels is shown in Figure 7-19 (58). The beneficial effect of delta ferrite is generally attributed to its interference with the propagation of cracks across the austenite matrix. To obtain a significantly improved resistance, however, considerable quantities of ferrite must be present, such as those found in duplex stainless steels. These are discussed in Section 7-2-4 which deals with duplex steels.

As noted in Chapter 2, plastic deformation of austenitic stainless steels results in both work hardening and partial transformation to alpha and epsilon martensites. Therefore, it is very difficult to separate effects due

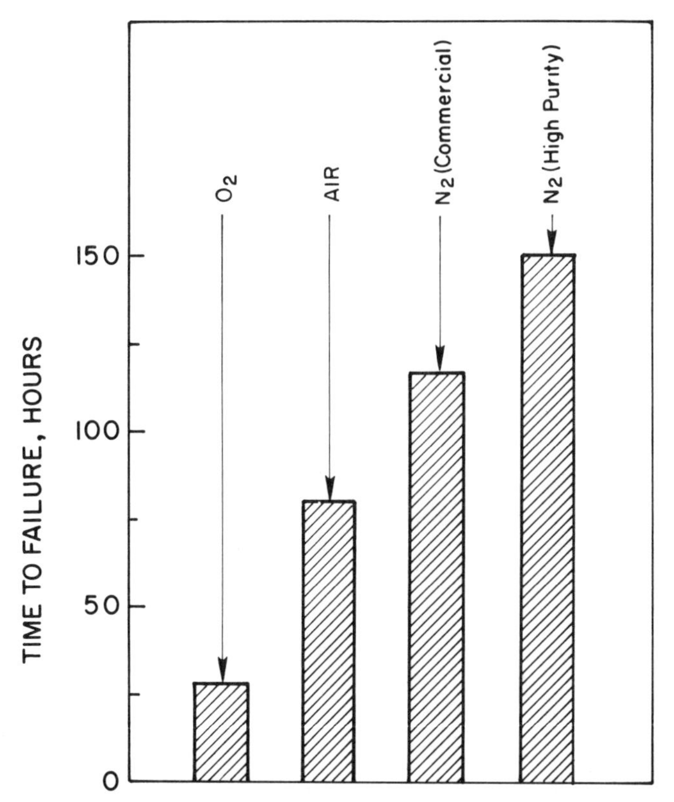

Figure 7-14. Effect of gas bubbling on the time to failure of type 304 stainless steel tested at 250 MPa in a 36.5% magnesium chloride solution at 110°C. (After Gräfen.)

to cold work (slip) from those due to transformation to martensites. Studies by Truman (59) using type 301, which readily transforms to martensites on cold working, show that at comparable applied stresses (e.g., 300 MPa) the cold worked and partially transformed material exhibits longer times to failure than the untransformed material, as shown in Figure 7-20. However, the deformed material has a higher yield stress. Studies by Cochran and Staehle (60), using prestrained wires of type 310, which shows little tendency for transformation to martensites, suggest a minimum at 10% prestrain, as indicated in Figure 7-21. Other reports of the observation of such minima have been compiled by Latanision and Staehle (19). The present state of understanding can be summarized as indicating that significant cold work (e.g., 35%) increases times to failure, as shown by the data of Truman (59) for type 321 and as illustrated in Figure 7-22.

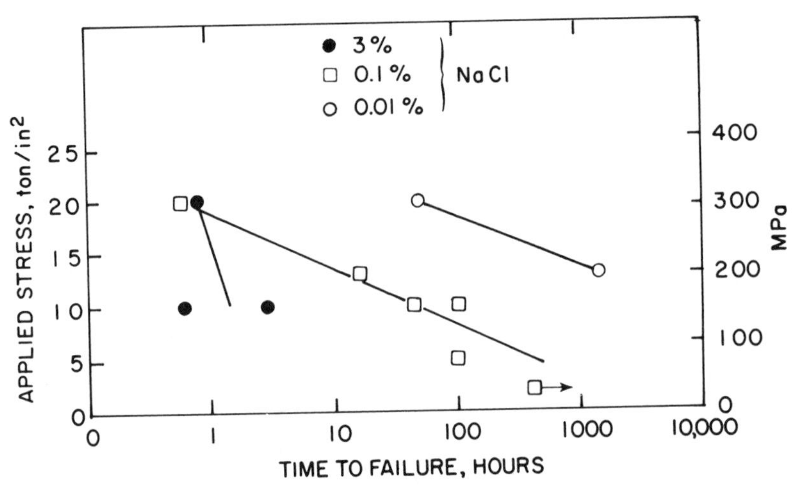

Figure 7-15. Effect of chloride concentration on the susceptibility to cracking of type 347 stainless steel in oxygen containing sodium chloride solutions at 250°C. (After Truman.)

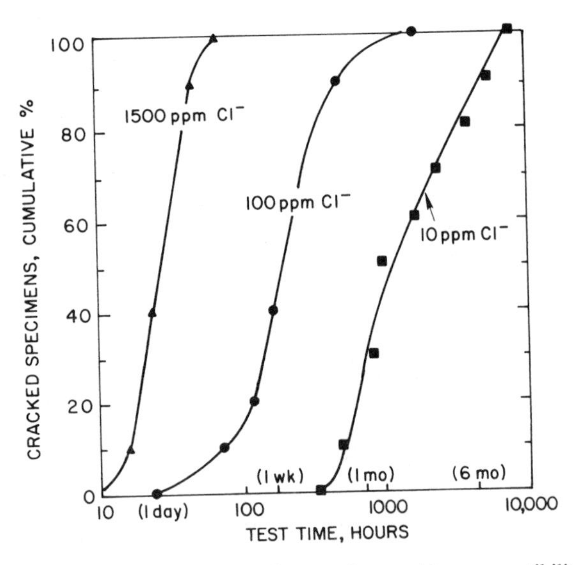

Figure 7-16. Effect of chloride concentration on the cracking susceptibility of type 304 stainless steel exposed at 100°C under the concentrating conditions of the Wick test. (After Warren.)

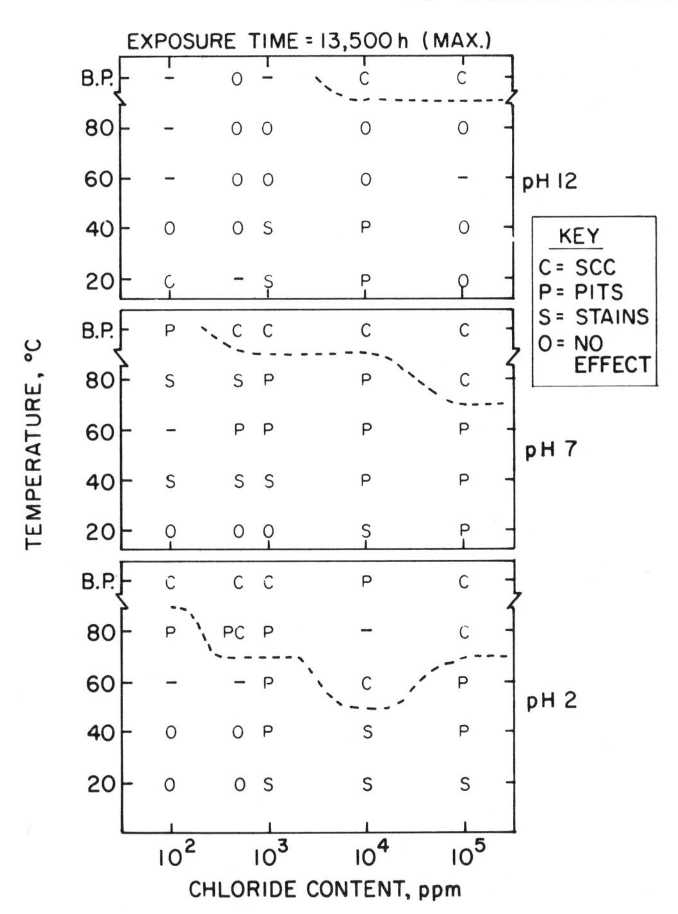

Figure 7-17. Effect of pH on the chloride content and temperature required to produce cracking of type 304 stainless steel in sodium chloride solutions. (After Truman.)

The effects of various surface finishes, which can produce variations due to local work hardening, martensitic transformation, residual stresses, embedded material from abrasives and machining equipment, or provide stress raisers, such as deep grooves or notches, are virtually impossible to quantify. All that can be said is that comparative evaluation of chloride cracking resistance should be carried out with nominally identical surface finishes. In particular, predictive laboratory evaluations should be carried out using surface finishes identical to those encountered in service.

Sensitization is detrimental to chloride cracking resistance. An example

TABLE 7-7. EFFECTS OF pH AND HEAT TREATMENT ON THE TIME TO FAILURE FOR SEVERAL AUSTENITIC STAINLESS STEELS IN BOILING 26% NaCl SOLUTION

pH	Alloy	Heat Treatment[1]	No. of U-Bend Specimens Tested	Time to Failure, Weeks Minimum	Time to Failure, Weeks Maximum
2	304	A	10	0.3	2.0
	304	A+L	10	1.0	1.0
	316	A	10	1.0	5.0
	316	A+L	10	0.7	2.3
	317	A	10	1.0	16.0
	317	A+L	10	1.0	8.0
4	304	A	4	NC	NC
	304	A+L	10	2.6	NC
	316	A	10	1.0	16.0
	316	A+L	10	1.0	11.0
	317	A	2	NC	NC
	317	A+L	2	NC	NC
6	304	A	2	NC	NC
	304	A+L	2	NC	NC
	316	A	2	NC	NC
	316	A+L	2	26[2]	NC
	317	A	2	NC	NC
	317	A+L	2	NC	NC
8	304	A	2	NC	NC
	304	A+L	2	NC	NC
	316	A	2	NC	NC
	316	A+L	2	NC	NC
	317	A	2	NC	NC
	317	A+L	2	NC	NC
10	304	A	2	NC	NC
	304	A+L	2	NC	NC
	316	A	2	NC	NC
	316	A+L	2	NC	NC
	317	A	2	NC	NC
	317	A+L	2	NC	NC
12	316	A	2	9.0[2]	NC
	316	A+L	2	NC	NC
	317	A	2	NC	NC
	317	A+L	2	NC	NC

[1]A = 27%CR + 1 h/1121°C/WQ
L = 1 h/677°C/WQ
[2]Crack in straight leg of U-bend away from bend.
NC = No cracking in 30-week test period.

of this, taken from the work of Truman (59) and summarized in Table 7-8, shows that type 304, when tested stressed in the Strauss solution (Table 6-1) and in boiling 40% calcium chloride, exhibits decreased resistance to both intergranular corrosion and chloride cracking with increased degree of sensitization. It should be noted that the chloride induced cracking in the sensitized material remained transgranular. Other studies (61) using

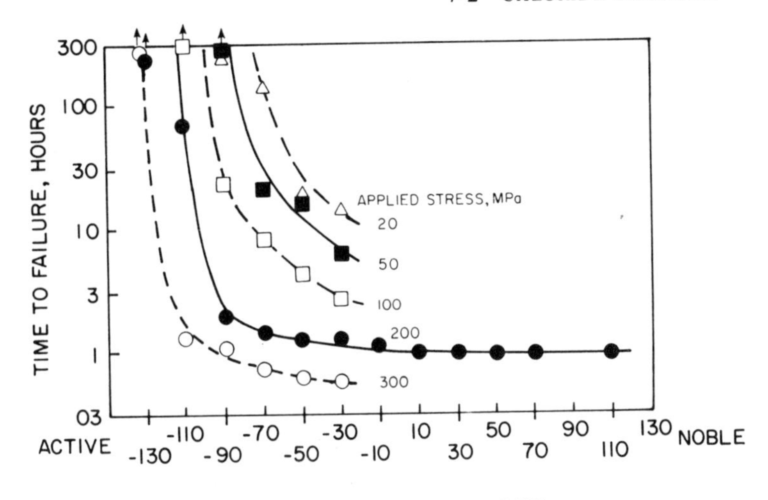

Figure 7-18. Relationship between applied potential, applied stress, and time to failure of solution annealed type 304 stainless steel in a 42% magnesium chloride solution at 144°C. (After Brauns and Ternes.)

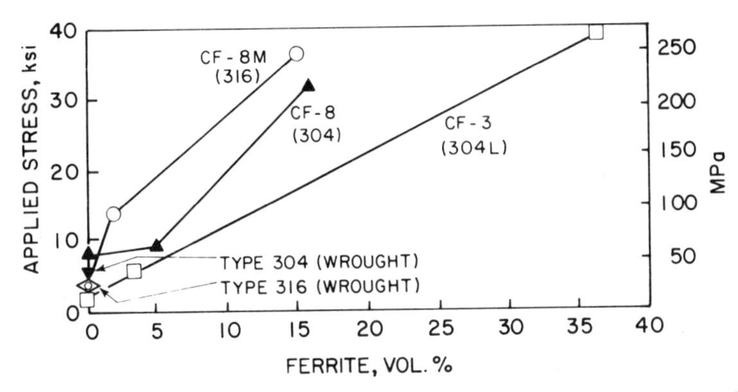

Figure 7-19. Effect of ferrite content on the stress required to induce chloride cracking in various cast stainless steels. Materials exposed for 8 hours in condensate from a 875 ppm chloride solution at 204°C. (After Fontana, Beck, and Flowers.)

high temperature sodium chloride environments containing oxygen have shown that cracking in type 304 changes from a transgranular to an intergranular type with sensitizing heat treatments. Studies by Kowaka and Kudo (17) have shown that in sensitized type 304 cracking is transgranular at high magnesium chloride concentrations (e.g., 45% $MgCl_2$) and predominantly intergranular at lower magnesium chloride concentrations (e.g., 20% $MgCl_2$).

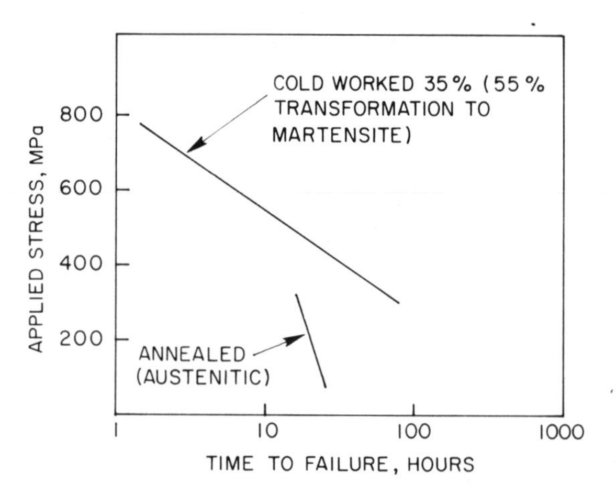

Figure 7-20. Effect of cold work on the susceptibility to cracking of type 301 stainless steel in a boiling 40% calcium chloride solution. (After Truman.)

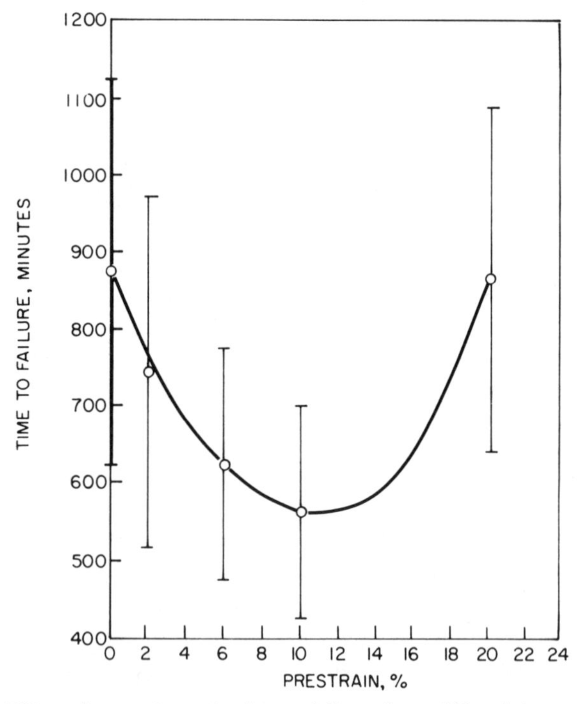

Figure 7-21. Effect of prestrain on the time to failure of type 310 stainless steel exposed to a magnesium chloride solution boiling at 154°C and stressed at 90% of the yield stress. (After Cochran and Staehle.)

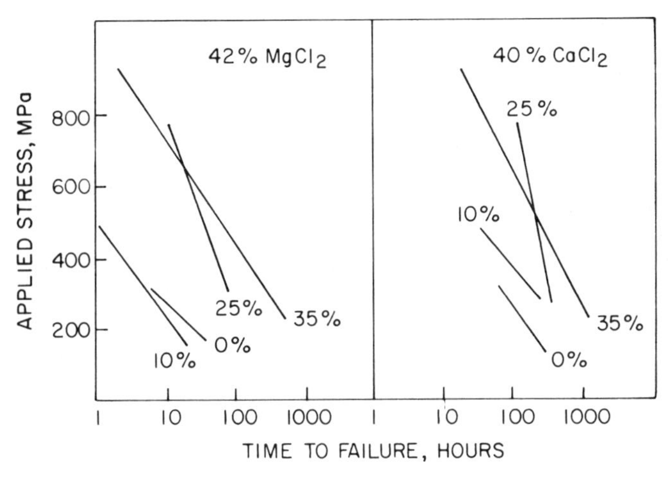

Figure 7-22. Effect of cold work (%) on the susceptibility to cracking of type 321 stainless steel in boiling magnesium chloride and calcium chloride solutions. (After Truman.)

TABLE 7-8. EFFECT OF SENSITIZATION ON THE TIME TO FAILURE OF TYPE 304 STAINLESS STEEL AT AN APPLIED STRESS OF 150 MPa

Corrodent	Time to Failure, hours 1050°C/AC	1050°C/AC + 2 h/650°C	1050°C/AC + 5 h/650°C
Boiling $CuSO_4$ + H_2SO_4 Solution	>500, NA	168, IGA	68, IGA
Boiling 40% $CaCl_2$ Solution	119, TC	74, TC	30, TC

AC = air cooled.
IGA = intergranular attack.
TC = transgranular cracking.
NA = no attack.

Regarding sigma, there appear to have been few attempts to evaluate its effect on chloride cracking resistance. A metallographic study by Warren (62) has shown that the transgranular cracks produced in boiling 42% magnesium chloride avoid or bypass sigma produced in high silicon types 316 and 317 by heating for 4 hours at 871°C.

Regarding sulfide inclusions, studies by Lang (63) and Warren (62) have failed to identify any relationship between sulfide stringers and the transgranular cracks produced in boiling magnesium chloride environments. The cracks appear to ignore the sulfide stringers. Other studies suggest that there may be an association between crack initiation and sulfides (64).

7-2-3 Ferritic Stainless Steels

Background. The ferritic stainless steels of the AISI 400 series usually exhibit general corrosion or pitting attack in boiling magnesium chloride environments rather than the formation of discrete cracks (49). However, the formation of definite cracks in chloride environments in the 400 series of stainless steels have been reported in several instances. For example, Bednar (65) has reported the cracking of type 434, type 430, and Fe-18% Cr-2% Mo ferritic stainless steels in lithium chloride solutions, Streicher (66) has demonstrated the cracking of sensitized type 446 in boiling magnesium chloride and sodium chloride solutions, and Brown (10) has described the cracking of type 430F in the marine atmosphere. Because of the low usage of ferritic stainless steels in the more demanding technological applications where chloride cracking can occur, this subject has not received much attention in the past. However, more recent laboratory studies, associated with the development of the high chromium, low interstitial ferritic stainless steels, have provided some insight into the factors governing their susceptibility to chloride cracking. For a discussion of the current understanding of the stress corrosion behavior of these developmental ferritic stainless steels the reader should consult the review by Steigerwald et al. (67).

Among factors that have so far been identified as being detrimental to the chloride cracking resistance of ferritic stainless steels are the presence of certain alloying elements, sensitization (induced by heat treatment or welding), cold work, high temperature embrittlement, and the precipitation of α' (475°C embrittlement). Because of the metallurgical complexity of ferritic stainless steels, it is not known whether all of these factors are related to the phenomenon identified as chloride cracking in austenitic stainless steels, or also represent manifestations of other phenomena, such as stress aided intergranular corrosion along sensitized grain boundaries and hydrogen embrittlement. However, for the purpose of the present discussion it is assumed that effects that can be detected in boiling magnesium chloride tests relate to chloride cracking.

Effects of Composition. Among alloying elements that have been identified as detrimental to the chloride cracking resistance of ferritic stainless steels are copper, nickel, molybdenum (in the presence of nickel), cobalt (in the presence of molybdenum), ruthenium, and carbon. There are also some indications that sulfur, in the alloy or sulfur containing hydrogen ion recombination poisons in the environment can facilitate cracking in chloride environments.

The copper and nickel levels that introduce susceptibility to cracking in magnesium chloride environments in annealed and in welded Fe-18%

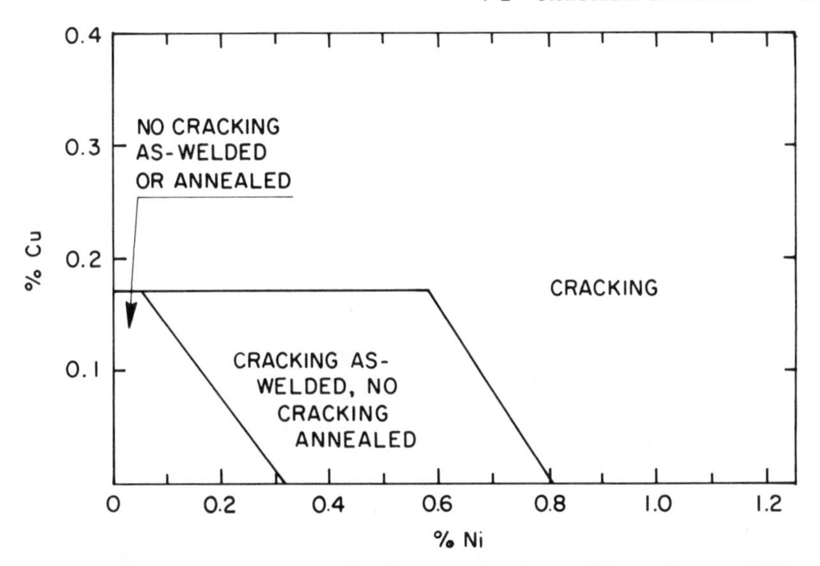

Figure 7-23. Effect of copper and nickel contents on the cracking resistance of U-bend specimens of ferritic Fe-18% Cr-2% Mo-0.35% Ti-0.015% C-0.015% N stainless steels exposed to a magnesium chloride solution boiling at 140°C. (After Steigerwald et al.)

Cr-2% Mo alloys are shown in Figure 7-23 (67). The detrimental effect of molybdenum in the presence of nickel is shown in Figure 7-24 (68). The detrimental effect of cobalt is noted in a study by Bond and Dundas (68). Since molybdenum can be an alloying addition, and copper, nickel, and cobalt can be present if scrap is used in the melt charge, these effects could become technologically significant. The reported detrimental effect of ruthenium (69) is of more academic interest, since significant amounts of this element are not generally found as an impurity in stainless steels. Regarding carbon, Streicher (66) has shown that for a 28.5% Cr-4.0% Mo stainless steel increasing the carbon content from 20 ppm to 171 ppm leads to cracking in a magnesium chloride solution boiling at 155°C. The suggestion that sulfur may be detrimental derives from the report (10) that type 430F (0.15% S, minimum) can exhibit cracking in the marine atmosphere. Also, Hoxie (41) has noted the cracking of 26% Cr-1% Mo low interstitial ferritic stainless steel in 132°C water containing chlorides, hydrogen sulfide, ammonia, and traces of oil, thiocyanates, and organic acids. The possibility that these sulfur related failures may be associated with hydrogen embrittlement cannot be ruled out, and a better insight would seem desirable.

In view of the importance of nitrogen on the sensitization behavior of

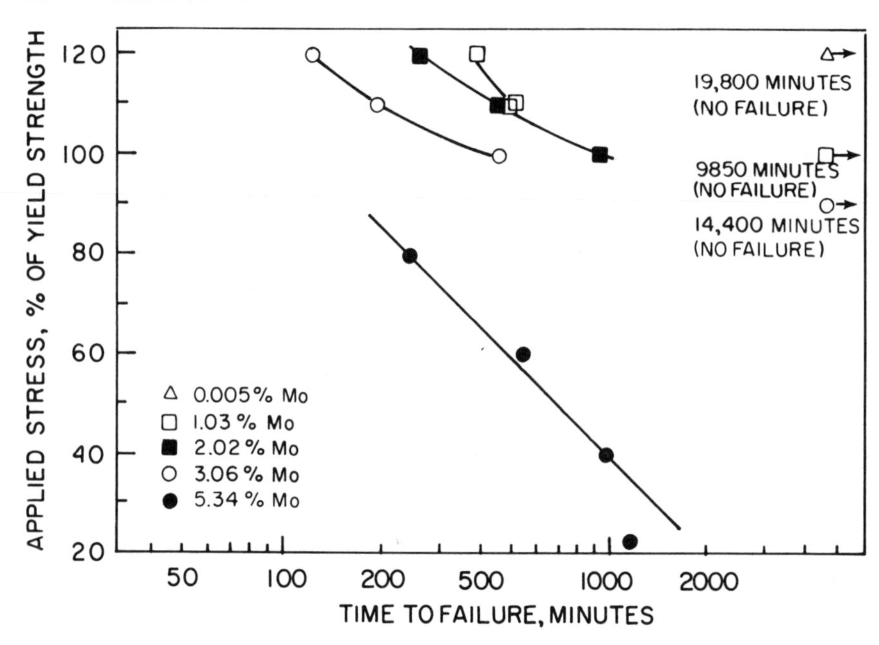

Figure 7-24. Effect of applied stress on the times to failure of tensile specimens of experimental ferritic Fe-17% Cr-1% Ni stainless steels containing various amounts of molybdenum in a magnesium chloride solution boiling at 140°C. (After Bond and Dundas.)

ferritic stainless steels, some studies of the role of this element in stress corrosion cracking would also seem desirable.

There have been few attempts to obtain a mechanistic understanding of the compositional factors affecting the chloride cracking susceptibility of ferritic stainless steels. However, Newberg and Uhlig (70) have shown that the addition of nickel to an 18% chromium stainless steel raises the corrosion potential, E_{corr} (defined in Chapter 3), in the noble direction. In terms of their advocated adsorption theory (see Table 7-1), which presupposes that a critical potential exists for the adsorption of chloride ions that induce cracking, they propose that it is the raising of E_{corr} above this critical potential that results in cracking.

Effects of Microstructure. It is probably true to say that the presence of structural features that reduce the ductility of ferritic stainless steels increase their susceptibility to chloride cracking. As discussed in Chapter 2, ductility is reduced by the precipitation of carbonitrides (high temperature embrittlement), α' (475°C embrittlement), and to some degree also by cold work. These three factors have been shown to increase suscepti-

TABLE 7-9. EFFECT OF VARIOUS HEAT TREATMENTS ON U-BENDS OF 26% Cr-2% Mo FERRITIC STAINLESS STEELS EXPOSED TO MAGNESIUM CHLORIDE SOLUTION BOILING AT 140°C

Composition, Wt. %			Cracking Time (Hours) After Indicated Heat Treatment Temperature							
Co	Cu	Ni	815°C	870°C	925°C	980°C	1040°C	1095°C	1150°C	TIG Weld
0.10	0.004	0.044	NF*	--	--	--	--	--	--	NF
0.32	0.004	0.044	NF	--	--	--	--	--	NF	NF
--	0.004	0.19	NF	--	--	--	--	--	--	NF
--	0.004	0.26	NF	--	--	--	--	--	NF	NF
--	0.004	0.36	NF	--	--	--	NF	NF	2-20	24-50
--	0.004	0.49	NF	NF	16-24	0-14	168-184	2-20	2-20	5-24
--	0.059	0.046	NF	--	--	--	--	--	--	NF
--	0.13	0.046	NF	--	--	--	--	--	NF	NF
--	0.22	0.046	NF	--	--	NF	NF	NF	144-168	52-68
--	0.28	0.064	NF	--	NF	16-24	28- 92	--	2-20	24-50

*NF = tests discontinued with no cracks detectable after 500 to 934 hours.

TABLE 7-10. EFFECT OF HEAT TREATMENT ON CRACKING INCIDENCE OF FERRITIC STAINLESS STEELS IN A 100 PPM CHLORIDE + 200 PPB OXYGEN SOLUTION AT 332°C. C-RING SPECIMENS STRESSED AT 90% OF THE YIELD STRENGTH

Material	Condition	Observation
Fe-18%Cr-2%Mo-0.5%Ti	Mill Annealed	NC
	3 h/475°C	NC
	20 h/475°C	C
E-BRITE 26-1	Mill Annealed	NC
	100 h/475°C	NC
	300 h/475°C	C

NC = no cracking in 10,578 hours.
C = intergranular cracking during 10,578 hours.

bility to chloride cracking. The detrimental effect on chloride cracking resistance of high temperature embrittlement is shown in Table 7-9 (67), of 475°C embrittlement in Table 7-10 (71), and of severe cold work in Table 7-11 (67).

There do not appear to have been any systematic investigations of the effect of sigma or of sulfide inclusions on chloride cracking resistance.

The detrimental effects of furnace sensitization on chloride cracking resistance of type 446 and welding induced sensitization of 28.5% Cr-4.2% Mo ferritic stainless steels containing small amounts of copper and nickel have been reported and discussed by Streicher (66, 72).

TABLE 7-11. RESULTS OF EXPOSURE OF U-BENDS OF 26% Cr-2% Mo FERRITIC STAINLESS STEELS IN MAGNESIUM CHLORIDE SOLUTION BOILING AT 140°C AS AFFECTED BY COLD WORK

Composition, Wt. %			Cracking Time (Hours) After Indicated Cold Work		
Co	Cu	Ni	30%	50%	75%
0.10	0.004	0.044	--	--	NF*
0.32	0.004	0.044	NF	72-96	342-358
--	0.004	0.19	NF	NF	98-142
--	0.004	0.26	NF	NF	6- 22
--	0.004	0.36	NF	6-22	4- 6
--	0.004	0.49	6-22	6-22	6- 22
--	0.059	0.046	--	--	NF
--	0.13	0.046	--	--	NF
--	0.22	0.046	NF	NF	46- 70
--	0.28	0.046	36-40	6-22	6- 22

*NF = tests discontinued with no cracks detectable after 600 to 888 hours.

7-2-4 Duplex Stainless Steels

Supplied in the annealed condition, duplex stainless steels have been used commercially, mainly as seamless tubing. In the annealed condition these steels are usually considered to be more resistant to chloride cracking than the common austenitic grades, such as types 304 and 316. The relative stress corrosion behavior of type 316 and the duplex stainless steel 3RE60 in a magnesium chloride solution is shown in Figure 7-25. Tests employing U-bend specimens of the microduplex IN-744 and austenitic type 304 in various less aggressive chloride environments have demonstrated the improved cracking resistance of the duplex steel, as summarized in Table 7-12 (73). Reported service experience with annealed 3RE60 tubing in industrial chloride solutions has been favorable (74).

However, welding is known to impair the stress corrosion resistance of IN-744. From a metallurgical viewpoint, welding will destroy the microduplex structure of the annealed material, creating continuous regions of ferrite in the heat affected zone, and, as discussed in the previous section, the precipitation of carbonitrides (high temperature embrittlement), the precipitation of α' (475°C embrittlement), and sensitization can increase the susceptibility of ferrite to chloride cracking. Significantly, cracking has been reported in the ferrite portion of the heat affected zone of welds in IN-744 when welded U-bend specimens were exposed in the

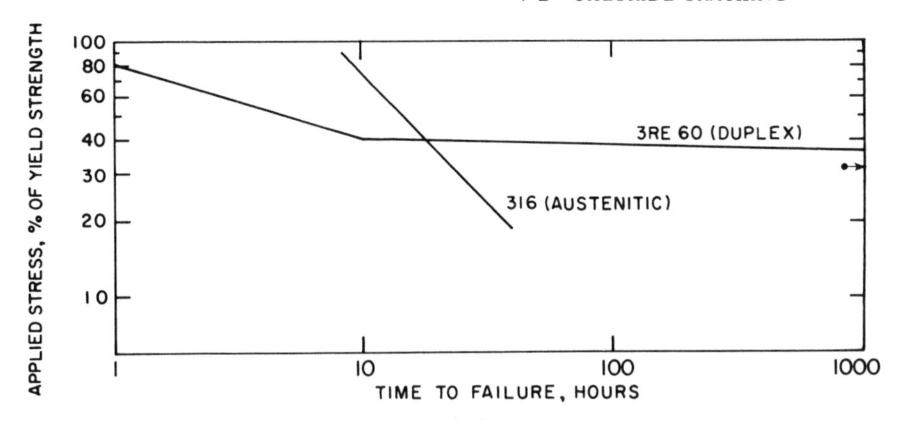

Figure 7-25. Effect of applied stress on the times to failure of type 316 and 3RE60 stainless steel tensile bars in a boiling 45% magnesium chloride solution. (Courtesy of Sandvikens Jernverks Aktiebolag.)

TABLE 7-12. RESULTS OF STRESS CORROSION TESTS USING U-BENDS OF ANNEALED IN-744 AND TYPE 304 STAINLESS STEELS

Test Medium	Time to Failure, days	
	IN-744	Type 304
36% $CaCl_2$, Boiling (110°C)	NF	5
3% NaCl, Vapor (95°C)	NF	4-6
28% NaCl, Vapor (95°C)	NF	3
3% NaCl, Solution (95°C)	NF	NF - 90
28% NaCl, Solution (95°C)	NF	NF
90% NaOH, Solution[a] (300°C)	NF[b]	14
42% $MgCl_2$, Solution (145°C)	1 - NF[c]	1

NF = no failure in 30 days.
a = autoclave tests.
NF[b] = no failure in 15 days.
NF[c] = no failure in 42 days.

vapor phase above a 3% sodium chloride solution at 95°C (75). Furthermore, Anderson and Novak (76) have reported cracking in the heat affected zone of welded IN-744 when exposed as U-bend specimens to the marine atmosphere and a chemical plant atmosphere contaminated with traces of phosphoric and sulfuric acids.

7-2-5 Martensitic and Precipitation Hardening Stainless Steels

As noted by Brown (10), the modern view, arising from extensive studies, is that most of the service failures involving high strength steels, which include high strength stainless steels, are due to hydrogen embrittlement. This is discussed in the last part of this chapter.

7-3 CAUSTIC CRACKING

7-3-1 General Background

The handling and containment of caustic soda (sodium hydroxide) does not pose a technological problem since adequate materials are readily available. For example, the low carbon grade of unalloyed nickel, Nickel 201 (0.02% C maximum) can handle caustic soda, without stress corrosion cracking, at all concentrations and temperatures of current technological interest (77). As noted by Fontana and Greene (78), general corrosion resistance in caustic is almost directly proportional to the nickel content of a given alloy, and the nickel containing austenitic stainless steels are used in some applications involving moderate temperature caustic at concentrations of 50% or less. However, it has been known for some time that at certain combinations of caustic concentration and temperature stainless steels can exhibit caustic cracking (79, 80).

More recently there has been an increased interest in this topic as a result of the use of austenitic alloys in heat exchanger systems of water cooled nuclear power plants. It is well known from conventional boiler technology that boiling and steam blanketing at heat transfer surfaces can give rise to very high local concentrations of caustic. The extent to which caustic can concentrate, at equilibrium conditions, as a function of the temperature difference between the bulk environment and a thin film of liquid at the heat transfer surfaces, is shown in Figure 7-26 (81).

Apart from the power industry, caustic cracking resistance also appears to be a topic of interest in the chemical, petrochemical, and pulp and paper industries, although few publications attest to this fact.

7-3-2 Austenitic Stainless Steels and Higher Alloys

A recent summary by Speidel (42) of available caustic cracking data for types 304, 347, 316, and 321 is shown in Figure 7-27. While there is some uncertainty about the position of the conservative line designated as "tentative safe SCC limit," the data show that there is an inherent danger of caustic cracking in strong caustic solutions when the temperature approaches 100°C. Sensitization appears to be detrimental at lower caustic concentrations. Figure 7-27 shows that at nuclear power plant steam generator temperatures of approximately 300°C these steels could exhibit rapid cracking. This is a fact well recognized by steam generator designers who utilize higher nickel alloys with improved caustic and chloride cracking resistance for these applications.

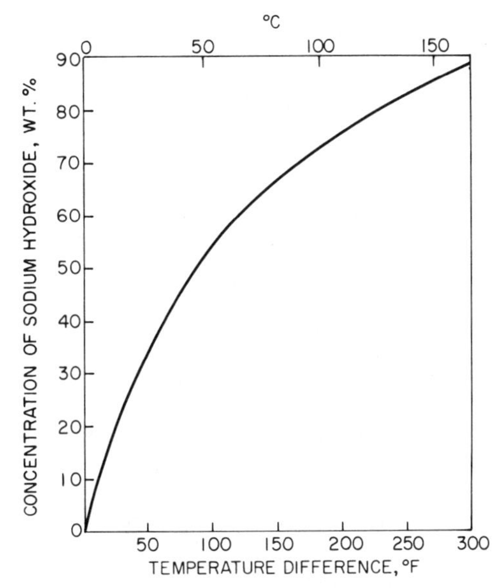

Figure 7-26. Caustic content attainable in a concentrating film of boiler water. (After Hecht et al.)

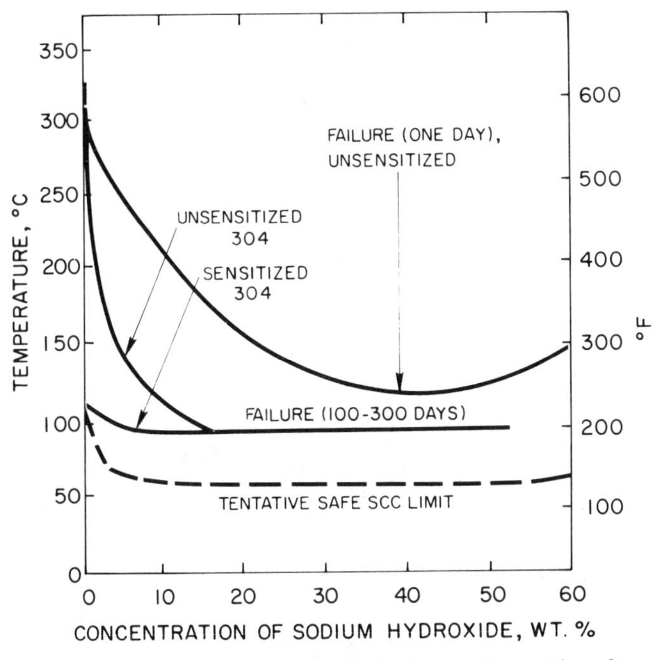

Figure 7-27. Temperature and concentration limits for caustic cracking of type 304, 347, 316, and 321 stainless steels. (After Speidel.)

In considering the factors affecting caustic cracking resistance, careful attention must be paid to the oxygen level of the caustic solution. While there is no evidence to suggest that oxygen is necessary to cause caustic cracking, major changes in the ranking of alloys can occur depending on whether oxygen is present or not. Recent studies (82, 83) have shown that in deaerated concentrated (50%) sodium hydroxide solutions cracking resistance increases with increasing nickel content of the alloy, with Nickel 201 exhibiting complete resistance, as shown in Figure 7-28 (82). The data shown in Figure 7-28 were obtained using U-bend specimens which measure the effects of both initiation and propagation of cracks. Tests using precracked (wedge-opening-load) fracture mechanics specimens have shown that the beneficial effect of nickel is also reflected in

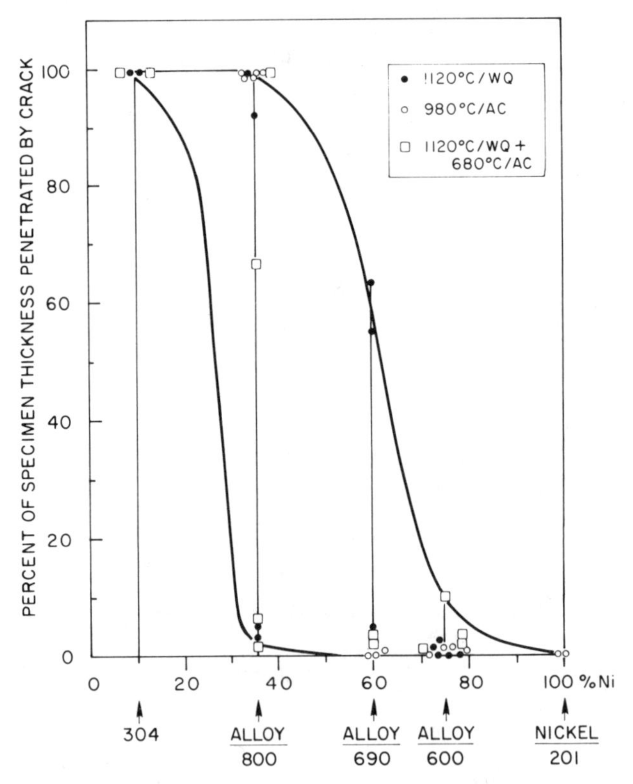

Figure 7-28. Caustic cracking behavior of U-bend specimens of various alloys in a deaerated 50% sodium hydroxide solution at 316°C. Exposure time = 5 weeks. (After Sedriks, Floreen, and McIlree.)

TABLE 7-13. THRESHOLD STRESS INTENSITIES IN DEAERATED 50% NaOH AT 316°C

Material*	Ni Content, Wt. %	Threshold Stress Intensity MPa $\sqrt{m}$	ksi $\sqrt{inch}$
Type 304	9	<1.1	<1
INCOLOY alloy 800	34	14.3	13
INCONEL alloy 690	60	22.0	20
INCONEL alloy 600	76	30.8	28

*Heat Treatments: Alloy 600, 1 h/1121°C/WQ
Other alloys, 1 h/1065°C/WQ

increasing threshold stress intensities, as summarized in Table 7-13 (82). The presence of molybdenum as an alloying element does not improve cracking resistance in deaerated 50% NaOH (83).

Figure 7-29 (11) shows that at lower caustic concentrations (e.g., 10% NaOH) the beneficial effect of nickel is less pronounced, with all three high nickel alloys (i.e., alloys 800, 690, and 600) exhibiting comparable cracking resistance.

In aerated (oxygen containing) solutions of sodium hydroxide high levels of both nickel and chromium appear to be necessary for resistance to caustic cracking, as indicated in Figures 7-30 (83) and 7-31 (84).

Figure 7-32 illustrates the observation by Truman (39) that in oxygenated caustic solutions times to failure increase with decreasing applied stress. This suggests that stress relief annealing could be a remedy. However, McIlree and Michels (83) have reported that annealing stressed U-bend specimens of type 304 stainless steel for 4 hours at 593°C did not prevent cracking on subsequent exposure to deaerated 50% sodium hydroxide at 284°C and 332°C. This, together with the observation that no meaningful threshold stress intensity could be determined for type 304 in this environment (Table 7-13), poses questions both as to the effectiveness of stress relief annealing for stainless steels in deaerated caustic solutions and as to the mechanism of cracking that could be operative.

The mitigation of caustic cracking by the addition of phosphate has received much more attention. Boiler water treatments employing phosphate additions to prevent the development of free caustic have been known for several decades and have been considered and used to prevent caustic cracking of stainless steels in boiler water (81, 85, 86). Chlorides (87) and chromates (88) have also been reported to inhibit the caustic cracking of austenitic stainless steels.

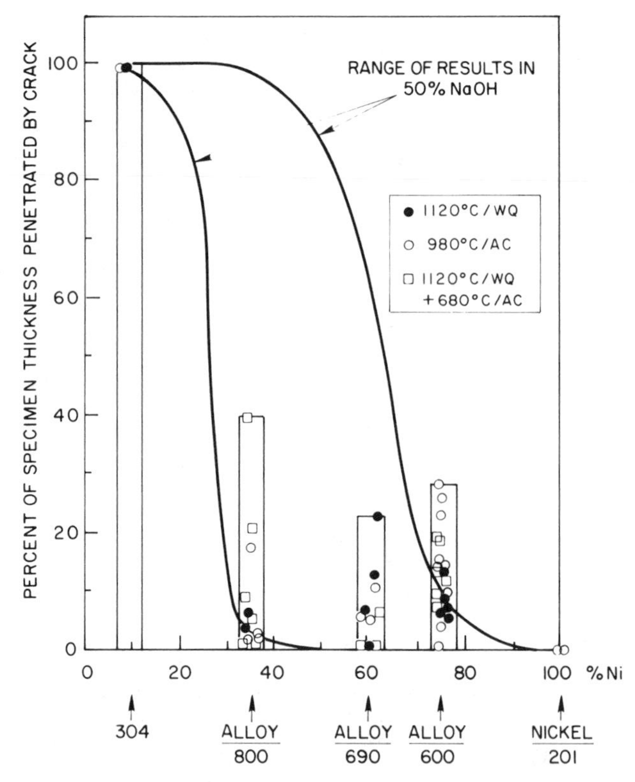

Figure 7-29. Caustic cracking behavior of U-bend specimens of various alloys in a deaerated 10% sodium hydroxide solution at 316°C. Exposure time = 6 weeks. (After Sedriks, Schultz, and Cordovi.)

7-3-3 Ferritic, Duplex, and Martensitic Stainless Steels

There appear to have been few evaluations of the caustic cracking resistance of stainless steels other than austenitics. Probably the most extensive investigation of commercial alloys to date has been that by Wilson, Pement, and Aspden (71) who examined the caustic cracking behavior of ferritic, duplex, and martensitic stainless steels under conditions insufficiently severe to crack type 304 (i.e., at 90% of the yield strength). Their data are shown in Table 7-14. Heat treatments designed to cause sensitization and 475°C embrittlement were detrimental to the caustic cracking resistance of the higher chromium ferritic stainless steels. The low chromium ferritic (type 405) exhibited heavy general corrosion. Heat treatments designed to produce 475°C embrittlement in the duplex stain-

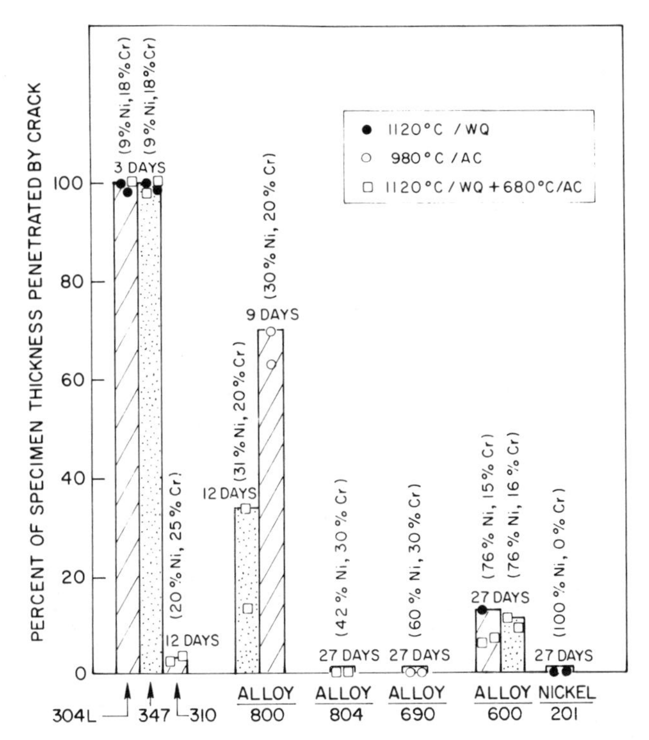

Figure 7-30. Caustic cracking behavior of U-bend specimens of various alloys in an aerated 50% sodium hydroxide solution at 300°C. (After McIlree and Michels.)

less steel were also detrimental. The martensitic stainless steel (type 410) exhibited cracking after tempering at 482°C, which gives a hardness of Rockwell C = 50.

McIlree and Michels (83) have reported cracking of type 446 and E-Brite 26-1, heat treated at 871°C, in deaerated 50% sodium hydroxide at 316°C, as summarized in Table 7-15, and Dillon (89) has noted the caustic cracking of the straight chromium (ferritic) grades of stainless steel. However, Bond et al. (90) reported uniform corrosion or intergranular attack in types 430 and 434 exposed to boiling 25% sodium hydroxide solution, as given in Table 7-16, rather than the formation of discrete cracks. No cracking was reported in U-bends of the annealed duplex stainless steel IN-744 after a 15 day exposure to 90% sodium hydroxide at 300°C (Table 7-12). More recent studies by Kowaka and Kudo (91) have shown that the introduction of the austenitic phase in a ferritic matrix

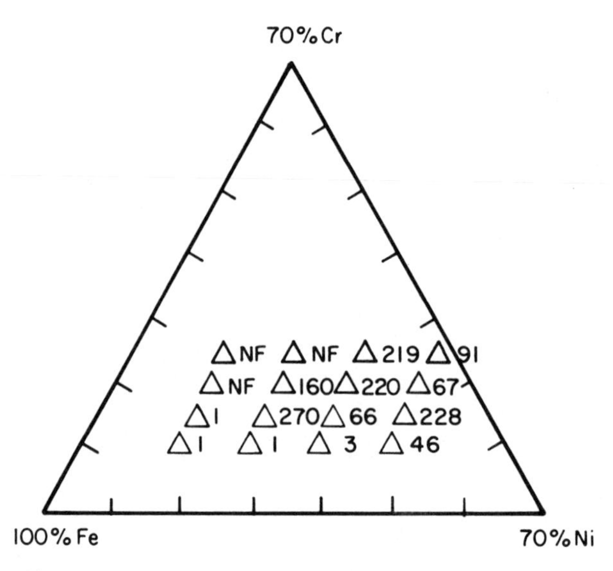

Figure 7-31. Times to failure (hours) of various Fe-Cr-Ni alloys in a 50% sodium hydroxide solution at 300°C with 14 MPa oxygen. Specimens stressed at 140 MPa. NF denotes no failure. (After Truman and Perry.)

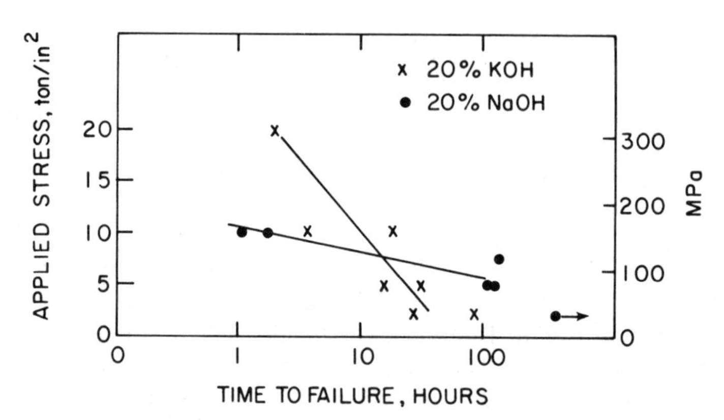

Figure 7-32. Effect of applied stress on the time to failure of type 347 stainless steel in oxygenated sodium hydroxide and potassium hydroxide solutions at 300°C. (After Truman.)

TABLE 7-14. CRACKING BEHAVIOR OF VARIOUS STAINLESS STEELS IN DEAERATED 10% NaOH SOLUTION AT 332°C. C-RING SPECIMENS STRESSED AT 90% OF THE YIELD STRENGTH

Stainless Steel	Condition	Observation
Type 304 (Austenitic)	Mill Annealed	NC
	10 h/649°C	NC
Type 405 (Ferritic)	Mill Annealed	NC, GC
	1 h/1010°C/FC	NC, GC
Fe-18%Cr-2%Mo-0.5%Ti (Ferritic)	Mill Annealed	NC, IGA
	1 h/1010°C/FC	C
	3 h/475°C	C
	20 h/475°C	C
E-BRITE 26-1 (Ferritic)	Mill Annealed	NC
	1 h/1010°C/FC	C
	100 h/475°C	NC
	300 h/475°C	C
3RE60 (Duplex)	Mill Annealed	NC, GC, IGA
	10 h/649°C	NC, GC, IGA
	100 h/475°C	C
	300 h/475°C	C
Type 410 (Martensitic)	Mill Annealed	NC, GC
	Tempered 649°C (Rc = 28)	NC, GC
	Tempered 565°C (Rc = 41)	NC, GC
	Tempered 482°C (Rc = 50)	C

NC = no cracking in 4800 hours.
C = cracking during 4800 hours.
GC = heavy general corrosion.
IGA = intergranular attack.
FC = furnace cooled.
Rc = hardness on Rockwell C scale.

TABLE 7-15. CRACKING BEHAVIOR OF FERRITIC STAINLESS STEELS IN DEAERATED 50% NaOH SOLUTION AT 316°C. U-BEND SPECIMENS

Material	Condition	Observation
Type 446	0.5 h/871°C/WQ	C, GC
E-BRITE 26-1	0.5 h/871°C/WQ	C, GC
	0.5 h/871°C/AC	C, GC
26-1S	0.5 h/871°C/AC	NC, GC

NC = no cracking in 10 weeks.
C = cracking during 10 weeks.
WQ = water quenched.
AC = air cooled.
GC = general corrosion.

TABLE 7-16. RESULTS OBSERVED AFTER EXPOSING 0.05 CM DIAMETER WIRES OF FERRITIC STAINLESS STEELS TO BOILING 25% NaOH SOLUTION AT A STRESS OF 365 MPa

Material	Condition	Exposure Time, hours	Observation
Type 430	Annealed	336	Uniform Corrosion
	0.25 h/982°C	428	Intergranular Attack
Type 434	Annealed	336	Uniform Corrosion
	1 h/482°C	355	Uniform Corrosion
	0.25 h/982°C	336	Intergranular Attack

TABLE 7-17. CAUSTIC CRACKING RESISTANCE OF VARIOUS EXPERIMENTAL STAINLESS STEELS TESTED AS BENT BEAM SPECIMENS IN BOILING 30% NaOH (119°C) FOR 300 HOURS

Alloy Composition, Wt. %					0.5 h/900°C/WQ		0.5 h/1100°C/WQ		0.16 h/1300°C/WQ	
Cr	Ni	Mo	C	N	Phases	Result	Phases	Result	Phases	Result
24.6	0.05	2.07	0.023	0.018	α	NF	α	C	α	C
24.9	0.50	2.11	0.029	0.020	α	NF	α	C	α	C
25.2	0.96	2.09	0.027	0.019	α	NF	α	C	α	C
24.9	1.95	2.09	0.030	0.018	α	NF	α	C	α	C
25.2	3.38	2.03	0.027	0.021	α+γ	NF	α	C	α	C
25.2	4.41	2.14	0.032	0.024	α+γ	NF	α+γ	C	α	C
24.9	5.92	2.09	0.027	0.017	--	--	α+γ	NF	α	C
24.8	7.63	2.10	0.024	0.014	--	--	α+γ	NF	α	NF
24.9	9.50	2.10	0.024	0.019	--	--	α+γ	NF	α	NF

NF = no failure.
C = intergranular caustic cracking.
α = ferrite
γ = austenite

improves the caustic cracking resistance of experimental 25% chromium stainless steels, as described in Table 7-17, although general corrosion resistance is decreased.

7-4 OXYGEN CRACKING

As noted in Figure 7-13, sensitized type 304 can exhibit cracking in high temperature-high oxygen solutions containing minimal amounts of chloride. Since chlorides can concentrate in crevices and pits, at first hand this type of attack would seem to be explainable as a form of chloride cracking, which can be accelerated by sensitization. However, studies by Copson and Economy (92) suggest that cracking in high temperature-high oxygen solutions should be regarded as having an identity separate from chloride (or caustic) cracking. In fact, their work suggests that it is the localized absence of oxygen that appears to cause cracking, since it is most readily observed in oxygen containing solutions in the presence of a crevice. As noted in Chapter 5, oxygen depletion can readily occur within a crevice.

The main reason for suggesting a mechanism other than chloride cracking derives from the observation that alloys resistant to chloride cracking can exhibit rapid cracking in the oxygen-plus-crevice situation, as illustrated in Figure 7-33(92). The data shown in Figure 7-33 were obtained using double U-bend specimens in which the crevice is formed between the inner and outer members of the double U-bend. Cracking initiates within the crevice and propagates into the outer surface of the inner U-bend. Cracking is intergranular, accelerated by sensitization, and occurs in types 304, 347, alloy 800, and in the chloride cracking resistant alloys 600 and 625. Of the austenitic Fe-Cr-Ni alloys, alloy 690 appears to be highly resistant to this form of cracking, as indicated in Table 7-18 (11).

Extensive studies by investigators at the General Electric Company have shown that the accelerating effect of a crevice is not necessary to obtain cracking in sensitized type 304, and they have defined numerous metallurgical, mechanical, and environmental parameters that affect this form of attack (93, 94). This topic is primarily of interest to boiling water nuclear reactor technology. However, even in these applications very few practical problems have been reported, and materials more resistant than sensitized type 304 appear to be readily available if needed (93).

7-5 CRACKING IN OTHER ENVIRONMENTS

In addition to polythionic acid cracking, discussed in Chapter 6, which is clearly a case of stress aided intergranular corrosion caused by sensitization, several other instances of cracking have been noted in certain environments. Acello and Greene (95) have noted the cracking of type 304 in sulfuric acid solutions containing sodium chloride, Spahn et al. (96) have reported the cracking of a copper containing austenitic stainless steel in boiling sulfuric acid solutions, and Theus and Cels (97) have detailed the cracking of sensitized type 304 in fluoride solutions.

For a compilation of stress corrosion failures in various complex industrial environments, generally containing chlorides, the reader should consult references 4 and 98.

7-6 HYDROGEN EMBRITTLEMENT

7-6-1 General Background

Hydrogen embrittlement is mainly a problem with high strength steels, and by far the greatest number of studies published have been concerned with high strength steels other than the ones belonging to the stainless

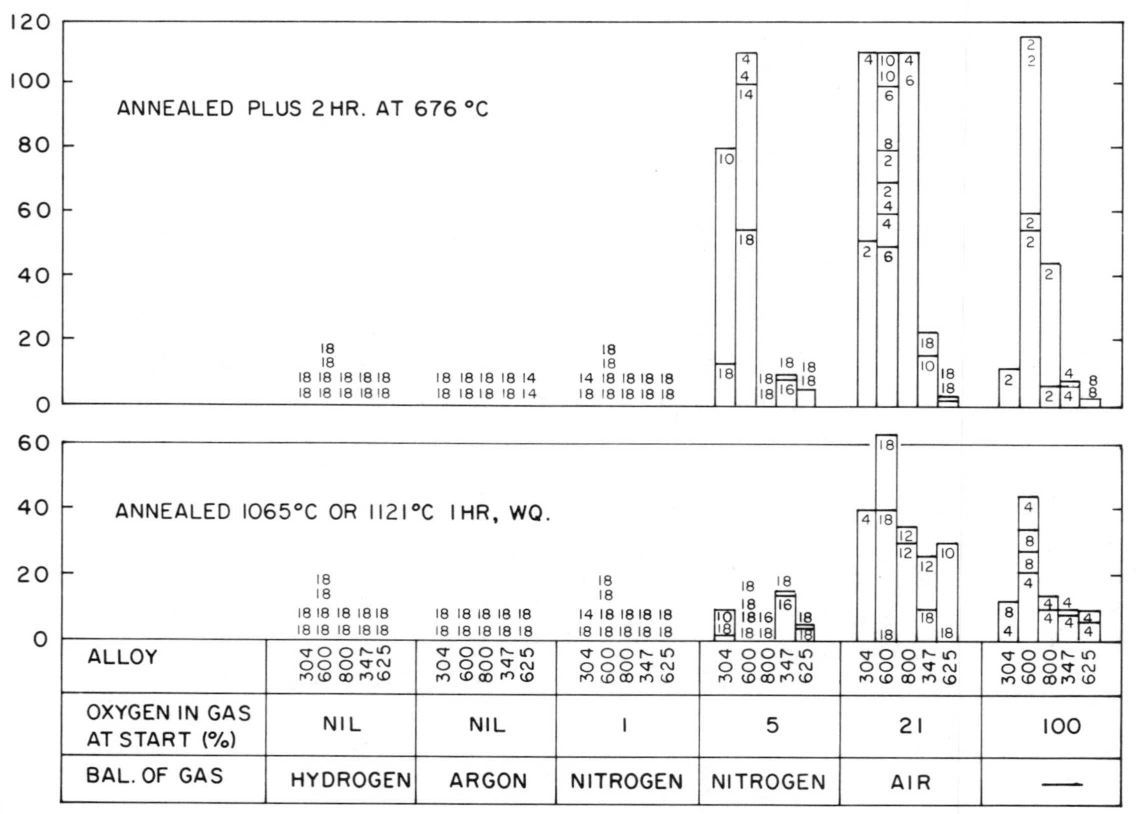

Figure 7-33. Comparison of stress corrosion behavior in crevice areas of double U-bend specimens of several alloys in 316°C high purity water adjusted to pH 10 with ammonium hydroxide at start-up. (After Copson and Economy.)

TABLE 7-18. STRESS CORROSION TESTS OF DOUBLE U-BEND SPECIMENS OF VARIOUS ALLOYS IN UNDEAERATED WATER AT 316°C TO EVALUATE THE OXYGEN-PLUS-CREVICE EFFECT

Alloy	Heat Treatment Code	Number of Specimens (Number of Heats)	Test Duration, Weeks	Number of Specimens Cracked/Number Destructively Examined	Maximum Depth of Attack, mm
Type 304	A_1+L_2	2 (1)	4	2/2	3.0
INCOLOY alloy 800	A_1+L_2	2 (1)	4	2/2	2.0
INCONEL alloy 600	A_1+L_2	2 (1)	2	2/2	2.6
INCONEL alloy 690	MA	2 (1)	48	0/2	0
	MA+L	2 (1)	48	0/2	0
	$MA+A_1$	2 (1)	48	0/2	0
	MA+CR	2 (1)	48	0/2	0
	A_1+L	2 (1)	48	0/2	0
	A_3+L	2 (1)	48	0/2	0
	A_2+W (12.7)	4 (1)	48	0/4	0
	A_2+W (25.4)	2 (1)	48	0/2	0
	MA+W (3.8)+L_1	2 (1)	48	0/2	0

Code: MA = mill annealed
CR = cold rolled 40%
A_1 = 1 h @ 1093°C/WQ
A_2 = 1 h @ 1150°C/WQ
A_3 = 1 h @ 1205°C/WQ
L = 2 h @ 650°C/AC
L_1 = 4 h @ 593°C/AC
L_2 = 5 h @ 704°C/AC
W (thickness, mm) = manual gas tungsten-arc welded at indicated thickness with matching filler.

group. Nevertheless, high strength stainless steels of the martensitic or precipitation hardening varieties have received a certain amount of attention, and for reviews of the state of understanding the reader is referred to the publications by Phelps (99), Poole (100), and Brown (10).

In considering service environments, it has been customary to discuss the cracking of high strength stainless steels in environments containing sulfides and in environments not containing sulfides. Sulfide environments are of particular interest to the oil industry, and there is a large measure of agreement with the view that cracking in sulfide environments (known as sulfide cracking) is a form of hydrogen embrittlement. Sulfide is a hydrogen ion recombination poison that promotes the entry of hydrogen into the metal.

For other aqueous environments, such as saline solutions, two cracking mechanisms have been discussed in the literature. For example, Phelps

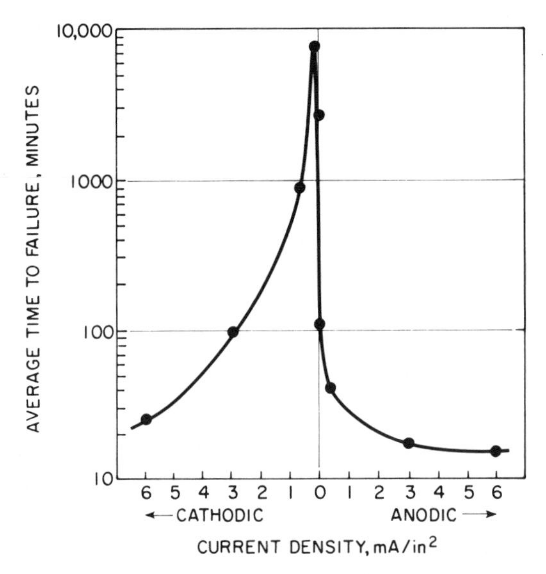

Figure 7-34. Effect of applied current on the time to failure of USS 12MoV stainless steel in an oxygenated 3% sodium chloride solution at pH 6.5. (After Bhatt and Phelps.)

(99) has made a distinction between cracking that is accelerated by the application of anodic currents and cracking that is accelerated by cathodic currents. An example of this behavior is shown in Figure 7-34 (101) for the case of a proprietary martensitic stainless steel, USS 12MoV (Fe-12% Cr-1% Mo-0.33% V-0.25% C), in a neutral salt solution. Subsequent studies, reviewed by Poole (100), suggest that in high strength steels hydrogen embrittlement is the crack propagation mechanism under both cathodic and anodic conditions. Anodic dissolution may, however, contribute to the initiation of the crack nucleus by processes such as pitting. It should also be noted that the presence of the chloride ion is not necessary to cause cracking in high strength stainless steels (99).

7-6-2 Cracking in Environments Not Containing Sulfides

Most evaluations of the cracking resistance of high strength stainless steels in sulfide free environments have been carried out either in natural marine atmospheres or in chloride solutions. Among parameters that determine resistance to cracking, yield strength has been identified as having a dominant effect. Cracking can occur quite readily in most high strength materials irrespective of composition or structure. This is illustrated by the data presented in Figure 7-35 (102) for a wide variety of high strength materials. It can be seen that martensitic and precipitation hard-

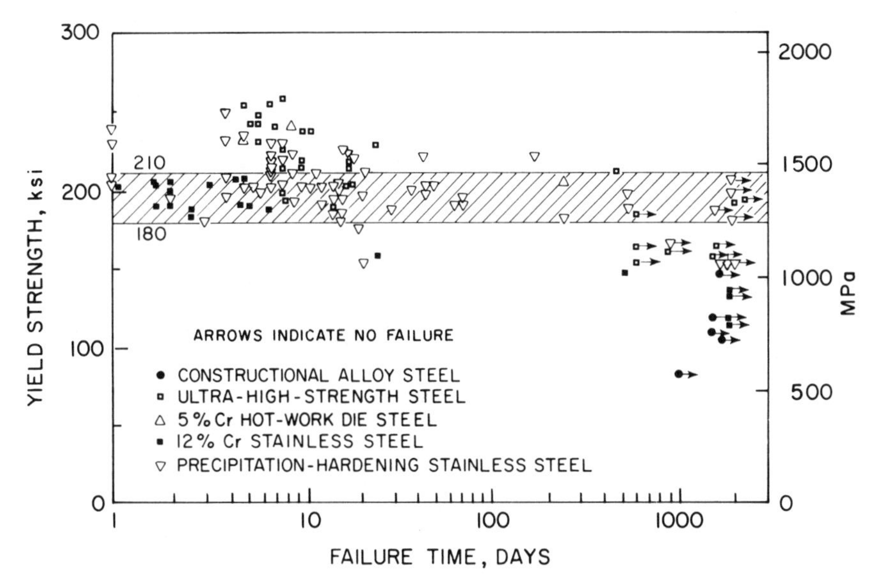

Figure 7-35. Effect of yield strength on the times to failure of various steels exposed to a marine atmosphere at 75% of the yield stress. (After Schmitt and Phelps.)

ening stainless steels can exhibit cracking in marine atmosphere at yield strengths above 1035 MPa (150 ksi). This applies to smooth specimens. In notched or precracked specimens cracking may occur at lower yield strengths.

Tempering, in the case of martensitic stainless steels, and overaging, in the case of precipitation hardening stainless steels, can significantly lower the yield strength and hence increase cracking resistance. An example of the effect of tempering on the yield strength and cracking resistance of a martensitic stainless steel in marine atmosphere is presented in Figure 7-36 (103). Accordingly, it has become accepted practice to define resistance to cracking in terms of yield strength, hardness, or heat treatment conditions. For example, the Marshall Space Flight Center (104) permits the use of precipitation hardening stainless steels in space vehicles and associated equipment only in certain conditions of heat treatment known to improve resistance to cracking, as outlined in Table 7-19 (104).

Bhatt and Phelps (101) showed that the pH of the environment can have an important effect on cracking resistance, as illustrated in Figure 7-37. For a martensitic stainless steel heat treated to give a yield strength of 1400 MPa (203 ksi), no cracking was observed at a pH higher than 11.5, whereas rapid cracking was found at a low pH. As pointed out by Brown

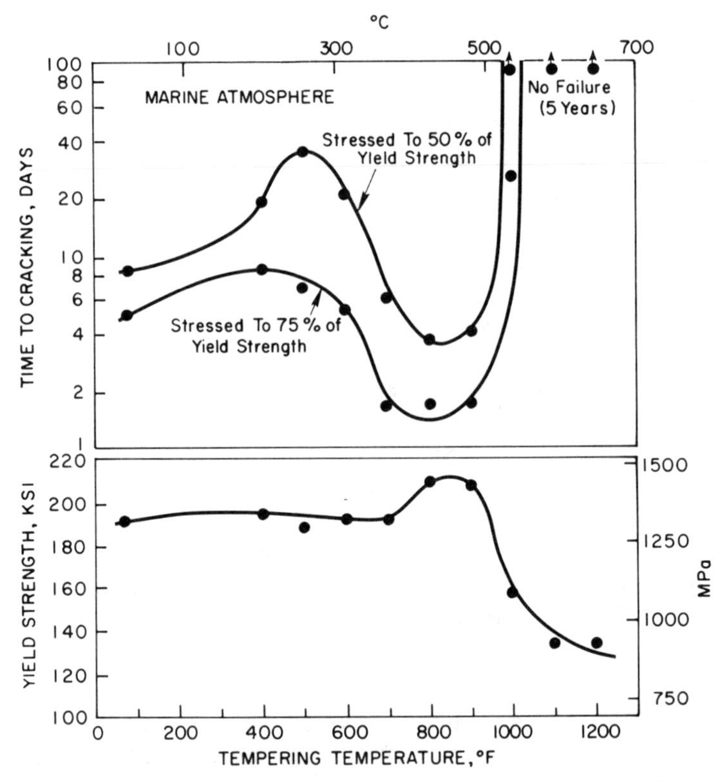

Figure 7-36. Effect of tempering temperature on the cracking resistance and yield strength of USS 12MoV stainless steel. (After Bates and Loginow.)

(10), acidification of the solution within crevices could lead to cracking not anticipated from tests in neutral solutions. The characterization of cracking resistance of high strength stainless steels in low pH solutions, typical of those developed within crevices, would seem to be an area worth further investigation.

There have been few attempts to explore the effects of microstructure and composition on cracking resistance. However, Lillys and Nehrenberg (105) have shown that the presence of delta ferrite minimizes the tendency for cracking in commercial martensitic stainless steels, and Truman (59) has noted the beneficial effects of columbium additions on the cracking resistance of certain experimental martensitic stainless steels. Fontana (106) has reported a correlation between crack initiation and sulfide inclusions for type 403 martensitic stainless steel.

TABLE 7-19. STAINLESS STEELS AND HIGHER ALLOYS REGARDED AS HAVING HIGH RESISTANCE TO HYDROGEN EMBRITTLEMENT BY MARSHALL SPACE FLIGHT CENTER

Alloy	Condition
AISI 300 Series Stainless Steel (Unsensitized)[1]	All
21-6-9 Stainless Steel	All
CARPENTER 20Cb	All
CARPENTER 20Cb3	All
A286	All
AM350	SCT 1000 and Above
AM355	SCT 1000 and Above
ALMAR 362	H1000 and Above
CUSTOM 455	H1000 and Above
15-5 PH	H1000 and Above
PH 14-8 Mo	CH900 and SRH950 and Above
PH 15-7 Mo	CH900
17-7 PH	CH900
NITRONIC 33[2]	All
HASTELLOY alloy C	All
HASTELLOY alloy X	All
INCOLOY alloy 800	All
INCOLOY alloy 901	All
INCOLOY alloy 903	All
INCONEL alloy 600[2]	Annealed
INCONEL alloy 625	Annealed
INCONEL alloy 718[2]	All
INCONEL alloy X-750	All
MONEL alloy K-500[2]	All
NI-SPAN-C alloy 902	All
RENE´ 41	All
UNITEMP 212	All
WASPALLOY	All

(1) Including weldments of 304L, 316L, 321 and 347.
(2) Including weldments.

7-6-3 Cracking in Environments Containing Sulfides

It is well known that the presence of hydrogen sulfide in various environments decreases the cracking resistance of high strength steels, including high strength stainless steels. This problem became prominent in the 1950s as a result of failures in the oil industry and has received considerable attention. As in the case of the sulfide free environments, most of the studies have dealt with steels other than high strength stainless steels, but there is sufficient information about the latter since many evaluations have included some of these materials.

The extensive investigation by Hudgins et al. (107), which included the precipitation hardening stainless steels AM-350, AM-355, A-286, and 17-4 PH, led to the conclusion that in sulfide containing environments none of the materials tested completely resisted cracking at yield

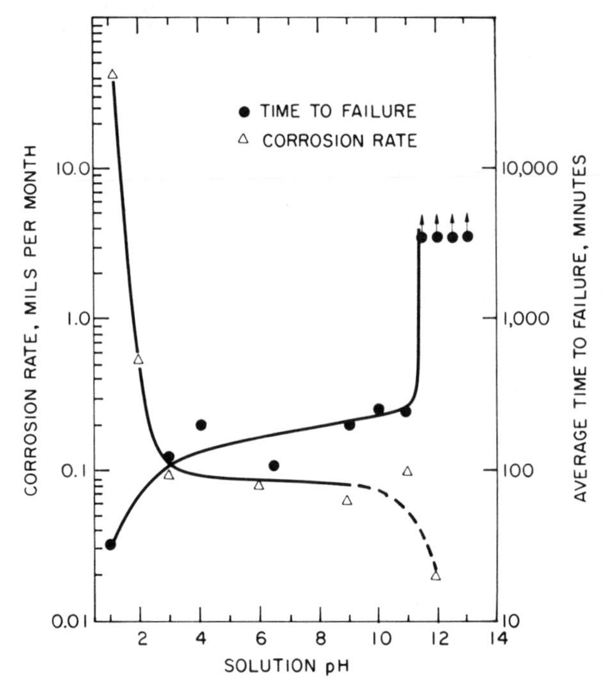

Figure 7-37. Effect of pH on the time to failure and initial corrosion rate of USS 12MoV stainless steel in an oxygenated 3% sodium chloride solution. (After Bhatt and Phelps.)

strengths above 690 MPa (100 ksi), except Monel alloy K-500. Hudgins et al. (107) also established that there exists a minimum hardness below which cracking does not occur for a given applied stress, and that this minimum hardness increases with decreasing hydrogen sulfide concentration. Numerous subsequent studies, covering a wide range of materials, have led to the issuance of NACE Standard MR-01-75 (108), which comprises a listing of materials for valves for production and pipeline service that are considered resistant to sulfide cracking. The part of the listing relating to stainless steels and higher alloys is summarized in Table 7-20. For a recent review of actual oil industry experience with high strength steels the reader should consult the publication by Treseder (109).

Factors, such as low pH, increase in temperature, galvanic coupling to anodic materials, and cold work facilitate cracking, and the NACE Standard MR-01-75, should be regarded as a guideline based on best current knowledge. The development of new alloys resistant to sulfide cracking

TABLE 7-20. STAINLESS STEELS AND HIGHER ALLOYS LISTED IN NACE STANDARD MR-01-75* AS ACCEPTABLE FOR SULFIDE ENVIRONMENTS

Stainless Steels				
Austenitic[a]	Ferritic[b]	Martensitic[c]	Precipitation Hardening	Higher Alloys
302	405	410	A286[d]	INCONEL alloy X-750[d]
304	430	CA15	17-4 PH[e]	HASTELLOY alloy C
304L		CA15M		MONEL alloy 400
310				INCONEL alloy 600[f]
316				MP35N[g]
316L				MONEL alloy K-500[f]
317				STELLITE Co-Cr-W Cast Alloys
321				CW-12M-1 (Cast)
347				CW-12M-2 (Cast)
CARPENTER 20Cb3				

[a]Annealed condition, not strengthened by cold work.
[b]Annealed condition, hardeness of Rockwell C = 22 (max.).
[c]Double tempered to hardness of Rockwell C = 22 (max.).
[d]Aged to hardness of Rockwell C = 35 (max.).
[e]Aged to hardness of Rockwell C = 33 (max.).
[f]Rockwell C = 35 (max.).
[g]Rockwell C = 50 (max.).

*Currently being revised. Reader should consult Technical Practices Committee of NACE for considered modifications. Some of the materials listed above may be susceptible to chloride cracking in certain environments.

should continue to be a subject of serious research activity, particularly as alloys with higher yield strengths become desirable for deep gas wells.

REFERENCES

1. A. J. Sedriks, *J. Inst. Met.*, Vol. 101, p. 225, 1973.
2. R. W. Staehle, *Proceedings of the NATO Conference on the Theory of Stress Corrosion Cracking in Alloys*, Ericeiria, Portugal, March 1971.
3. E. Escalante and W. F. Gerhold, *Galvanic and Pitting Corrosion—Field and Laboratory Studies*, ASTM STP-576, American Society for Testing and Materials, 1976, p. 81.
4. *Report on Stress Corrosion Cracking of Austenitic Chromium-Nickel Stainless Steels*, ASTM STP-264, American Society for Testing and Materials, 1960.
5. M. R. Louthan, *Corrosion*, Vol. 21, p. 288, 1965.
6. D. Tromans and J. Nutting, *Corrosion*, Vol. 21, p. 143, 1965.
7. F. A. Champion, *Symposium on Internal Stresses in Metals and Alloys*, Institute of Metals, London, 1948, p. 468.
8. R. N. Parkins, *Met. Rev.*, Vol. 9, p. 201, 1964.
9. A. J. Sedriks, P. W. Slattery, and E. N. Pugh, *Proceedings of the Conference on Fundamental Aspects of Stress Corrosion Cracking*, NACE, Houston, Tex., 1969, p. 673.
10. B. F. Brown, *Stress Corrosion Cracking Control Measures*, NBS Monograph 156, 1977, p. 55.
11. A. J. Sedriks, J. W. Schultz, and M. A. Cordovi, *Corrosion Engineering (Boshoku Gijutsu)*, Japan Society of Corrosion Engineering, Vol. 28, p. 82, 1979.
12. *Proceedings of the Conference on Fundamental Aspects of Stress Corrosion Cracking*, NACE, Houston, Tex., 1969.
13. *Proceedings of the Conference on Stress Corrosion Cracking and Hydrogen Embrittlement of Iron-Base Alloys*, NACE, Houston, Tex., 1977.
14. M. A. Scheil, *Symposium on Stress Corrosion Cracking of Metals*, ASTM STP-64, American Society for Testing and Materials, 1945, p. 395.
15. M. A. Streicher and A. J. Sweet, *Corrosion*, Vol. 25, p. 1, 1969.
16. I. B. Casale, *Corrosion*, Vol. 23, p. 314, 1967.
17. M. Kowaka and T. Kudo, *Trans. Jap. Inst. Met.*, Vol. 16, p. 385, 1975.
18. "Stress Corrosion Testing Methods," in *Stress Corrosion Testing*, ASTM STP-425, American Society for Testing and Materials, 1967, p. 3.
19. R. M. Latanision and R. W. Staehle, *Proceedings of the Conference on Fundamental Aspects of Stress Corrosion Cracking*, NACE, Houston, Tex., 1969, p. 214; G. J. Theus and R. W. Staehle, Reference 13.
20. H. R. Copson, *Physical Metallurgy of Stress Corrosion Fracture*, Interscience, New York, 1959, p. 247.
21. R. W. Staehle, *Trans. Inst. Chem. Eng.*, Vol. 47, p. T-227, 1969.
22. E. E. Denhard and R. R. Gaugh, *Stress Corrosion Testing*, ASTM STP-425, American Society for Testing and Materials, 1967, p. 41.
23. A. J. Sedriks, *Corrosion*, Vol. 31, p. 339, 1975.
24. R. L. Beauchamp, M. S. thesis, The Ohio State University, 1963.
25. J. J. Royuela and R. W. Staehle, reference 19.
26. H. H. Uhlig and R. A. White, *Trans. ASM*, Vol. 52, p. 830, 1960.
27. J. Hochman and J. Bourrat, *C. R. Acad. Sci.*, Vol. 255, December 17, 1962.
28. J. E. Truman, Brown-Firth Research Laboratories, private communication.

29. R. W. Staehle, J. J. Royuela, T. L. Raredon, E. Serrate, C. R. Morin, and R. V. Farrar, *Corrosion*, Vol. 26, p. 451, 1970.
30. F. S. Lang, *Corrosion*, Vol. 18, p. 378t, 1962.
31. D. G. Tufanov, *Metalloved. Term. Okrabotka Met.*, 15, April 1964.
32. D. van Rooyen, *Proceedings of the First International Conference on Stress Corrosion Cracking*, Butterworths, London, 1961.
33. J. G. Hines and E. R. W. Jones, *Corros. Sci.*, Vol. 1, p. 88, 1961.
34. M. da Cunha Belo and J. Montuelle, *Proceedings of the NATO Conference on the Theory of Stress Corrosion Cracking in Alloys*, Ericeiria, Portugal, March 1971.
35. H. Spähn, G. H. Wagner, and U. Steinhoff, "Methods to Prevent Stress Corrosion Cracking," *Techinscher Ueberwach.*, Vol. 14, p. 292, 1973.
36. H. Lefakis and W. Rostoker, *Corrosion*, Vol. 33, p. 179, 1977.
37. M. D. Speidel, *Corrosion*, Vol. 33, p. 199, 1977.
38. K. Fässler, in *Korrosionum*, H. Gräfen, F. Kahl, and A. Rahmel, Eds., Vol. 1, Verlag Chemie GmbH, Weinheim/Bergstrasse, Germany, 1974, p. 136.
39. J. E. Truman, *Methods Available for Avoiding SCC of Austenitic Stainless Steels in Potentially Dangerous Environments*, Stainless Steels, ISI Publication 117, The Iron and Steel Institute, London, 1969, p. 101.
40. K. L. Money and W. W. Kirk, *Mater. Performance*, Vol. 17, No. 7, p. 28, July 1978.
41. E. C. Hoxie, "Some Corrosion Considerations in the Selection of Stainless Steel for Pressure Vessels and Piping," in *Pressure Vessels and Piping: A Decade of Progress*, Vol. 3, The American Society of Mechanical Engineers, New York, 1977.
42. M. O. Speidel, *Handbook on Stress Corrosion Cracking and Corrosion Fatigue*, to be published by Advanced Research Projects Agency, Washington, D.C.
43. H. Gräfen, *Corros. Sci.*, Vol. 7, p. 177, 1967.
44. D. Warren, "Chloride-Bearing Cooling Water and the Stress Corrosion Cracking of Austenitic Stainless Steel," *Proceedings of the 15th Annual Purdue Industrial Waste Conference*, Purdue University, May 1960, p. 1.
45. A. W. Dana and W. B. DeLong, *Corrosion*, Vol. 12, p. 309t, 1956.
46. H. R. Baker, M. C. Bloom, R. N. Bolster, and C. R. Singleberry, *Corrosion*, Vol. 26, p. 420, 1970.
47. J. E. Truman, *Corros. Sci.*, Vol. 17, p. 737, 1977.
48. J. L. Nelson, The International Nickel Company, Inc., to be published.
49. A. S. Couper, *Mater. Prot.*, Vol. 8, p. 17, 1969.
50. E. Brauns and H. Ternes, *Werkst. Korros.*, Vol. 19, p. 1, 1968.
51. H. H. Uhlig, *J. Electrochem. Soc.*, Vol. 116, p. 173, 1969.
52. J. H. Pillips and W. J. Singely, *Corrosion*, Vol. 15, p. 450t, 1959.
53. S. P. Rideout, W. C. Rion, and J. Wade, *Trans. Am. Nucl. Soc.*, Vol. 7, p. 419, 1964.
54. M. Fujii and M. Kumada, *J. Jap. Inst. Met.*, Vol. 35, p. 560, 1971.
55. M. L. Holzworth and A. E. Symonds, *Corrosion*, Vol. 25, p. 287, 1969.
56. P. P. Snowden, *Nucl. Eng.*, Vol. 6, p. 409, 1961.
57. S. A. Balezin and N. I. Podovaev, *J. Appl. Chem. USSR*, Vol. 33, p. 1287, June 1960.
58. M. G. Fontana, F. H. Beck, and J. W. Flowers, *Met. Prog.*, Vol. 80, No. 6, p. 99, 1961.
59. J. E. Truman, "The Effects of Composition and of Structure on the Resistance to Stress Corrosion Cracking of Stainless Steels," *British Nuclear Energy Society Symposium on Effects of Environment on Material Properties in Nuclear Systems*, Paper no. 10, Institute of Civil Engineers, London, July 1971.
60. R. W. Cochran and R. W. Staehle, *Corrosion*, Vol. 11, p. 369, 1968.
61. L. R. Scharfstein and W. F. Brindley, *Corrosion*, Vol. 14, p. 60, 1958.
62. D. Warren, "Microstructure and Corrosion Resistance of Austenitic Stainless Steels,"

Sixth Annual Liberty Bell Corrosion Course, NACE, Philadelphia, Pa., September 1968.

63. F. S. Lang, *Corrosion,* Vol. 18, p. 378t, 1962.
64. R. F. Overman, *Corrosion,* Vol. 22, p. 48, 1966.
65. L. Bednar, *Corrosion,* Vol. 35, p. 96, 1979.
66. M. A. Streicher, *Stress Corrosion of Ferritic Stainless Steels,* Paper No. 68, presented at Corrosion/75, Toronto, Canada, April 1975.
67. R. F. Steigerwald, A. P. Bond, H. J. Dundas, and E. A. Lizlovs, *Corrosion,* Vol. 33, p. 279, 1977.
68. A. P. Bond and H. J. Dundas, *Corrosion,* Vol. 24, p. 344, 1968.
69. M. A. Streicher, *Platinum Met. Rev.,* Vol. 21, p. 51, 1977.
70. R. T. Newberg and H. H. Uhlig, *J. Electrochem. Soc.,* Vol. 119, p. 981, 1972.
71. I. L. W. Wilson, F. W. Pement, and R. G. Aspden, *Stress Corrosion Studies on Some Stainless Steels in Elevated Temperature Aqueous Environments,* Paper No. 136, paper presented at NACE Corrosion/77, San Francisco, Cal., March 1977.
72. M. A. Streicher, *Corrosion,* Vol. 30, p. 77, 1974.
73. S. N. Anant Narayan and G. N. Flint, *Recent Advances in Nickel-Containing Corrosion Resistant Alloys,* to be published in *Indian Chem. Manuf.*
74. R. F. Steigerwald, *Mater. Performance,* p. 9, September 1974.
75. *Preliminary Data: IN-744 Stainless Steel,* The International Nickel Company, Inc., October 1, 1969.
76. D. B. Anderson and C. J. Novak, *Stress Corrosion Cracking of High Strength Stainless and Low Alloy Steels in Chemical Plant Atmospheres,* Paper No. 152, presented at NACE Corrosion/76, Houston, Tex., 1976.
77. *Corrosion Resistance of Nickel and Nickel-Containing Alloys in Caustic Soda and Other Alkalies,* Corrosion Engineering Bulletin CEB-2, The International Nickel Company, Inc., New York, 1976.
78. M. G. Fontana and N. D. Greene, *Corrosion Engineering,* McGraw-Hill Book Company, New York, 1967.
79. H. Nathorst, Welding Research Council Bulletin No. 6, October 1950, p. 6.
80. A. K. Agrawal and R. W. Staehle, *Stress Corrosion Cracking of Fe-Cr-Ni Alloys in Caustic Environments,* Report No. C00-2018-21(Q6), Ohio State University, Columbus, Ohio, April-July 1970.
81. M. Hecht, E. P. Partridge, W. C. Schraeder, and S. F. Hall, in *The Corrosion Handbook,* H. H. Uhlig, Ed., John Wiley & Sons, Inc., New York, 1963, p. 520.
82. A. J. Sedriks, S. Floreen, and A. R. McIlree, *Corrosion,* Vol. 32, p. 157, 1976.
83. A. R. McIlree and H. T. Michels, *Corrosion,* Vol. 33, p. 60, 1977.
84. J. E. Truman and R. Perry, *Br. Corros. J.,* Vol. 1, p. 60, 1966.
85. G. C. Wheeler and E. Howells, "A Look at Caustic Stress Corrosion," *Power,* p. 86, September 1960.
86. E. Howells, "Caustic Stress Corrosion Tests on Stainless Steel," *Corros. Technol.,* p. 368, 1960.
87. A. V. Ryabchenkov, V. I. Gerasimov and V. P. Sidorov, *Prot. Met.,* Vol. 2, p. 217, May-June 1966.
88. R. W. Staehle and A. K. Agrawal, *Corrosion, Stress Corrosion Cracking, and Electrochemistry of the Fe and Ni Base Alloys in Caustic Environments,* Report to ERDA, Contract E (11-1)-2421, The Ohio State University, 1976.
89. C. P. Dillon, *Mater. Prot. Performance,* Vol. 11, p. 48, June 1972.
90. A. P. Bond, J. D. Marshall, and H. J. Dundas, "Resistance of Ferritic Stainless Steels to Stress Corrosion Cracking," *Stress Corrosion Testing,* ASTM STP-425, American Society for Testing and Materials, Philadelphia, Pa., 1967, p. 116.

91. M. Kowaka and T. Kudo, *Stress Corrosion Cracking Behavior of Ferritic and Duplex Stainless Steels in a Caustic Solution,* paper presented at the Spring Meeting of the Japan Society of Corrosion Engineering, Tokyo, Japan, May 1977.
92. H. R. Copson and G. Economy, *Corrosion,* Vol. 24, p. 55, 1968.
93. W. L. Clarke and G. M. Gordon, *Corrosion,* Vol. 29, p. 1, 1973.
94. R. L. Cowan and G. M. Gordon, *Proceedings of the International Conference on Stress Corrosion Cracking and Hydrogen Embrittlement of Iron-Base Alloys,* NACE, Houston, Tex., 1977, p. 1023.
95. S. J. Acello and N. D. Greene, *Corrosion,* Vol. 18, p. 322t, 1962.
96. H. Spähn, G. H. Wagner, and U. Steinhoff, *Proceedings of the International Conference on Stress Corrosion Cracking and Hydrogen Embrittlement of Iron-Base Alloys,* NACE, Houston, Tex., 1977, p. 80.
97. G. J. Theus and J. R. Cels, "Fluoride Induced Intergranular Stress Corrosion Cracking of Sensitized Stainless Steel," *Corrosion Problems in Energy Conversion and Generation,* The Electrochemical Society, Princeton, N.J., 1974, p. 384.
98. A. W. Loginow, J. F. Bates, and W. L. Mathay, *Mater. Prot. Performance,* Vol. 11, No. 5, p. 35, 1972.
99. E. H. Phelps, *Proceedings of the Conference on Fundamental Aspects of Stress Corrosion Cracking,* NACE, Houston, Tex., 1969, p. 398.
100. R. Poole, in *Corrosion,* L. L. Shreir, Ed., Vol. 1, Newnes-Butterworths, Boston, 1976, p. 8.62.
101. H. J. Bhatt and E. H. Phelps, *Corrosion,* Vol. 17, p. 430t, 1961.
102. R. J. Schmitt and E. H. Phelps, *J. Met.,* p. 47, March 1970.
103. J. F. Bates and A. W. Loginow, *Corrosion,* Vol. 20, p. 189t, 1964.
104. D. B. Franklin, *Design Criteria for Controlling Stress Corrosion Cracking,* Marshall Space Flight Center Document MSFC-SPEC-522A, November 1977.
105. P. Lillys and A. E. Nehrenberg, *Trans. ASM,* Vol. 48, p. 327, 1956.
106. M. G. Fontana, *Stress Corrosion Cracking in Type 403 Stainless Steel,* WADC Technical Report 56-242, ASTIA Document No. AD-97215, August 1956.
107. C. M. Hudgins, R. L. McGlasson, P. Mehdizadeh, and W. M. Rosborough, *Corrosion,* Vol. 22, p. 238, 1966.
108. *Mater. Performance,* Vol. 14, appendix, April 1975.
109. R. S. Treseder, *Proceedings of the International Conference on Stress Corrosion Cracking and Hydrogen Embrittlement of Iron-Base Alloys,* NACE, Houston, Tex., 1977, p. 147.

8

CORROSION FATIGUE, GALVANIC CORROSION, EROSION-CORROSION AND CAVITATION DAMAGE

8-1 CORROSION FATIGUE

8-1-1 Introduction

Fatigue is the term used to describe the cracking of a metal under repeated cyclic stresses. Generally fatigue cracks initiate and propagate at stresses below the yield strength after numerous cyclic applications of stress. As in the case of stress corrosion cracking, the fatigue crack propagates until the load bearing section of a material is reduced to the point at which the ultimate tensile strength is exceeded. At that point the material will separate by overload fracture. Laboratory fatigue tests are often conducted by subjecting a material to cyclic stresses of various amplitudes and measuring the time to failure. The nomenclature used in fatigue testing is shown in Figure 8-1 (1). More recently, fatigue tests have been developed that employ precracked fracture mechanics specimens and measure crack growth rate as a function of change in stress intensity. For an overview of such studies the reader should consult the publication by Hoeppner (1).

The term corrosion fatigue is used to describe fatigue cracking in the presence of liquid or gaseous environments that may or may not be particularly corrosive to the unstressed material. A convenient way of describing the effect of environment is by comparing the variation of fatigue fracture stress with the number of stress cycles applied (S-N curves) in air and in the environment of interest. An example is shown in Figure 8-2 (2). Curve *a* in Figure 8-2 shows the fatigue behavior of a titanium containing type 316 in air, with a 10^7 cycle fatigue limit* of 285

*Sometimes referred to as the endurance limit, particularly in British literature.

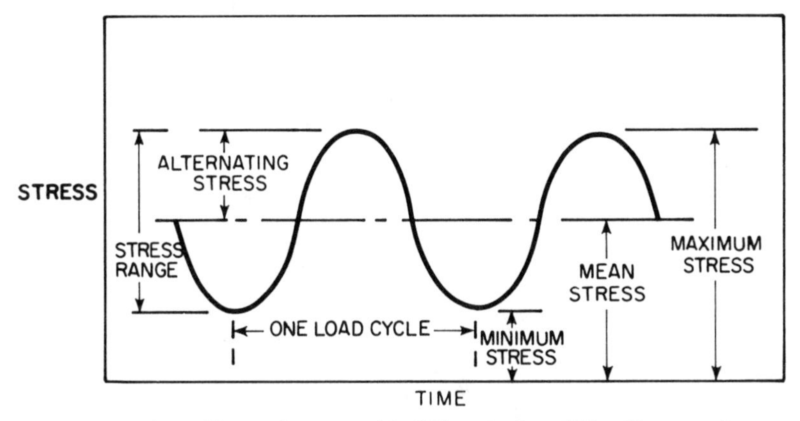

Figure 8-1. Nomenclature used in fatigue testing. (After Hoeppner.)

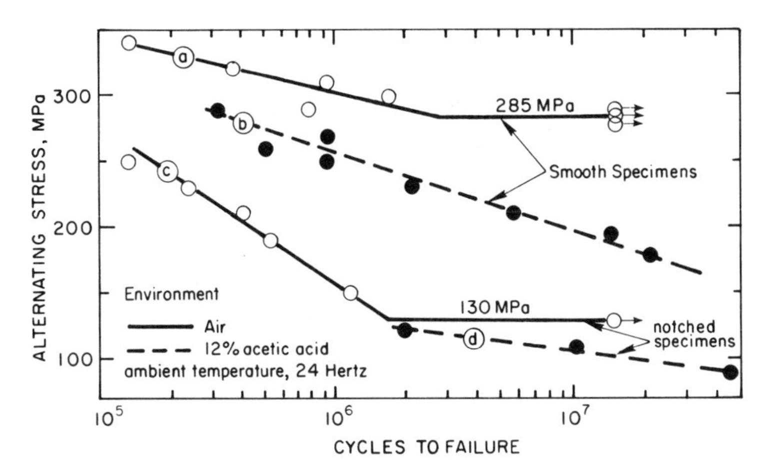

Figure 8-2. Fatigue behavior of rotating bend specimens of a titanium stabilized type 316 stainless steel in air and in acetic acid. (After Fässler.)

MPa. The same test in 12% acetic acid, curve *b* in Figure 8-2, shows cracking at lower stresses with no evidence of a fatigue limit. Rather than try to establish a corrosion fatigue limit by long and often costly extended tests, it has been the practice in alloy evaluations to compare the corrosion fatigue resistance of various alloys in terms of the stress at which no failure occurs after 10^7 or 10^8 cycles. This stress is often referred to as the corrosion fatigue strength (CFS) after a specified number of cycles (e.g., CFS at 10^8 cycles), and it should not be equated with or identified as the

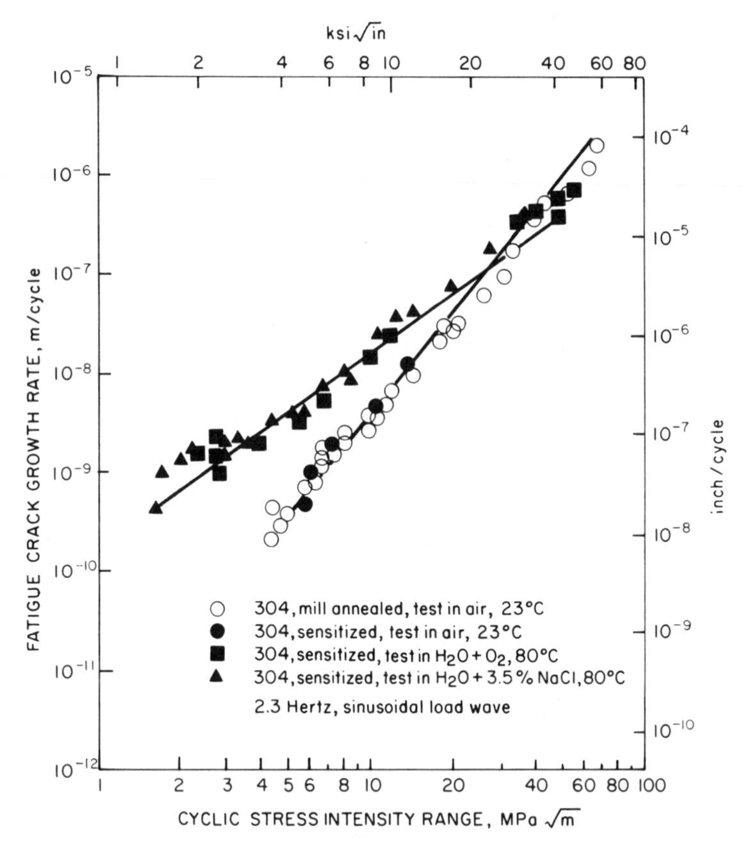

Figure 8-3. Effect of environment and cyclic stress intensity range on the growth rate of fatigue cracks in type 304 stainless steel. (After Speidel.)

corrosion fatigue limit, since it is not certain whether in some alloy/environment systems there is corrosion fatigue limit.

If the specimen used in the fatigue test contains a notch, which acts as a stress raiser, the fatigue limit in air is reduced (cf. curves *a* and *c* in Figure 8-2). The stresses at which cracking occurs in the environment are also reduced by the presence of a notch (cf. curves *b* and *d* in Figure 8-2). Pitting, intergranular corrosion, and stress corrosion can also produce stress raising surface defects that facilitate the initiation of fatigue cracks. However, it is now well established that the presence of a corrosive environment can also accelerate the propagation of fatigue cracks. The accelerating effect of a warm aqueous solution on the fatigue crack propagation in sensitized type 304 is shown in Figure 8-3 (3).

The mechanism of corrosion fatigue in stainless steels is a subject that has received little discussion. However, Spähn (4) has attempted to rationalize it in terms of the film rupture mechanism, proposed for stress corrosion cracking, modified to include a cyclic rupture and healing of the passive film. Corrosion fatigue crack path is usually transgranular, but can become intergranular in sensitized type 304.*

The technological importance of corrosion fatigue as a failure mode of stainless steels is difficult to assess. Table 1-3 indicates corrosion fatigue to be of relatively minor importance among corrosion failures encountered by DuPont. However, as noted by Collins and Monack (5) in their DuPont survey, fatigue failures accounted for one-third of the failures attributed to mechanical, as opposed to corrosion related, causes. It is conceivable that corrosion, or corrosion fatigue, may have contributed to the failures identified as deriving from mechanical fatigue.

8-1-2 Effects of Metallurgical Variables

Studies reported by Sedriks and Money (6) and May (7) of various stainless steels and higher alloys in flowing seawater suggest that ultimate tensile strength, pitting resistance, and grain size are parameters that affect corrosion fatigue strength. In general, increasing tensile strength, increasing pitting resistance (i.e., increasing nickel, chromium, and molybdenum contents), and decreasing grain size favor increased resistance to corrosion fatigue in seawater. However, none of these three factors have been studied systematically as a single variable with all other factors kept constant, so the findings should be regarded as tentative.

For stainless steels and higher alloys the most important variable appears to be tensile strength, as shown in Figure 8-4 (6). It should, of course, be recognized that high tensile strength materials not resistant to corrosion in seawater will not exhibit high corrosion fatigue strength. For alloys that do not pit in seawater, that is, Hastelloy alloy C, Inconel alloy 625, and IN-120,† there appears to be an almost linear relationship between CFS at 10^8 cycles and ultimate tensile strength. There is little separation between the behavior of types 304, 304L, 316, and 316L, all of which exhibit pitting in low velocity seawater, as illustrated in Figure 8-4. Prevention of pitting by cathodic protection at -0.85 V (SCE) has been found to raise the CFS at 10^8 cycles of CF-4 (cast type 304 with 0.04% C maximum) from a value of 62 MPa in the unprotected state to a value of 138 MPa in the protected state (7). Cathodic protection has also been shown to raise the CFS of the high strength maraging steels in seawater (8).

*T. Shoji et al., *Corrosion*, Vol. 34, p. 366, 1978.

†INCO Experimental alloy, Ni-21% Cr-14% Co-4% Mo-2.5% Ti-2% (Cb + Ta).

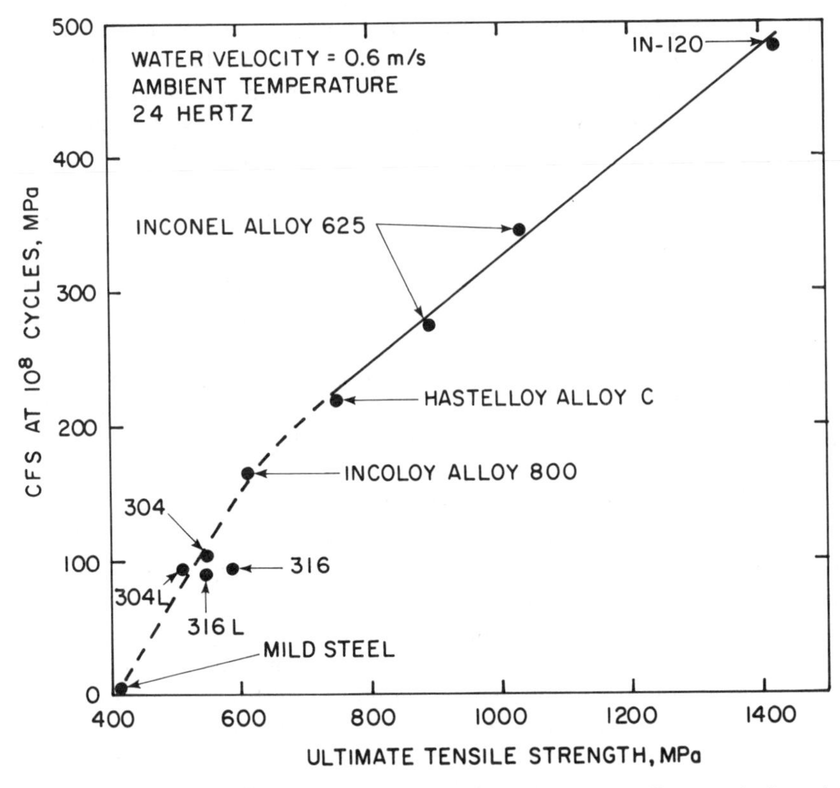

Figure 8-4. Variation of corrosion fatigue strength with ultimate tensile strength of rotating bend specimens of various alloys in seawater. (After Sedriks and Money.)

Increasing the tensile strength by precipitation hardening will also raise CFS. For example, May (7) has reported that 17-4 PH in the H900 condition exhibits a CFS at 10^8 cycles of 276 MPa in seawater. However, the Marshall Space Flight Center (9) regards such a high strength condition as marginal with respect to hydrogen embrittlement resistance and will not permit its use in space vehicles without prior approval and even then will restrict it to applications not involving high sustained tensile stress.

Strengthening produced by the introduction of delta ferrite in the microstructure, as in the case of duplex stainless steels, also increases CFS. For example, under the test conditions described in Figure 8-4, the duplex stainless steel IN-744 exhibits a CFS at 10^8 cycles of 172 MPa, which is a value significantly higher than that obtained for the austenitic stainless steels of the AISI 300 series. Similar results have been observed for various European duplex stainless steels, as summarized by Speidel (3). In the case of IN-744, the small grain size of the microduplex structure probably also increases CFS.

Strengthening produced by cold work and by nitrogen additions in type 316L stainless steel has been evaluated by Hughes, Jordan, and Orman (10) in a saline solution, 0.17 *M* NaCl, pH = 7.4, simulating a body fluid. While strengthening by cold work resulted in a significant increase in CFS at 10^8 cycles, the high nitrogen type 316L exhibited poor fatigue properties both in air and in the salt solution when tested in the presence of a notch, as reported in Table 8-1 (10). Regarding tests employing smooth (unnotched) specimens, Spähn (4) has reported that the CFS in synthetic seawater of certain German high nitrogen stainless steels is higher than the guaranteed minimum yield strength of type 347 produced in Germany. However, in view of the data shown in Table 8-1, the effect on CFS of strengthening by nitrogen additions, particularly in the presence of notches, is a subject deserving further study.

It is well known that decreasing grain size will result in an increasing fatigue limit in air. It would be expected, therefore, that a similar effect would be observed in the case of corrosion fatigue. While no systematic studies in this area have been reported for stainless steels, studies using Inconel alloy 718 (Ni-19% Cr-17% Fe-5% Cb-3% Mo-0.8% Ti-0.6% Al) suggest that the grain size effect is in evidence for corrosion fatigue strength in seawater, as shown in Figure 8-5 (6). The materials with grain sizes of 0.01 and 0.068 mm, which were tested in seawater, had the same ultimate tensile strength (1303 MPa) and very likely the same pitting resistance.

Sensitizing heat treatments have been shown to have a detrimental effect on CFS, both in the case of austenitic and martensitic stainless steels, as indicated in Figures 8-6 and 8-7, respectively. With regard to the

TABLE 8-1. EFFECTS OF STRENGTHENING BY 20% COLD WORK AND NITROGEN ADDITION ON THE FATIGUE PROPERTIES OF TYPE 316L STAINLESS STEEL IN AIR AND IN A 0.17*M* NaCl SOLUTION, pH = 7.5, AT AMBIENT TEMPERATURE. FREQUENCY = 100 HZ. AMSLER VIBROPHORE PUSH-PULL FATIGUE TESTS

	Fatigue Strength at 10^8 Cycles, MPa			
	Smooth Specimens		Notched Specimens	
Material and Treatment	Air	Salt Solution	Air	Salt Solution
Type 316L, 0.5 h/1050°C	241	203	103	76
Type 316L, 0.5 h/1050°C + 20% CW	441	338	124	107
Type 316L + 0.2%N, 0.5 h/1050°C	269	203	72	69

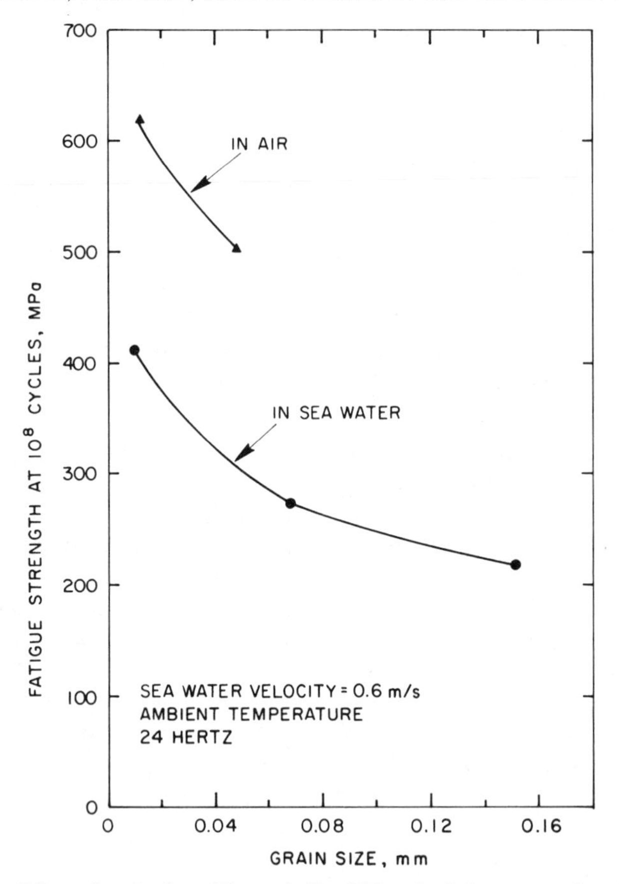

Figure 8-5. Effect of grain size of Inconel alloy 718 on its fatigue strength at 10^8 cycles in air and in seawater. (After Sedriks and Money.)

latter, tempering at 550°C is thought to cause chromium depletion in the vicinity of precipitated carbides (see Chapter 2).

8-1-3 Effects of Test and Environmental Variables

The effects on CFS of notches and cathodic protection have been noted in the preceding section. Among other test and environmental variables that have received some study for stainless steels are the effect of frequency of stressing and the pH of the solution. In the case of type 316L in a salt solution, it has been shown that decreasing the frequency leads to a small decrease in CFS at 10^7 cycles, as illustrated in Figure 8-8 (10). Figure 8-9 (10) shows that lowering the pH to 3 or below leads to a 50% reduction of

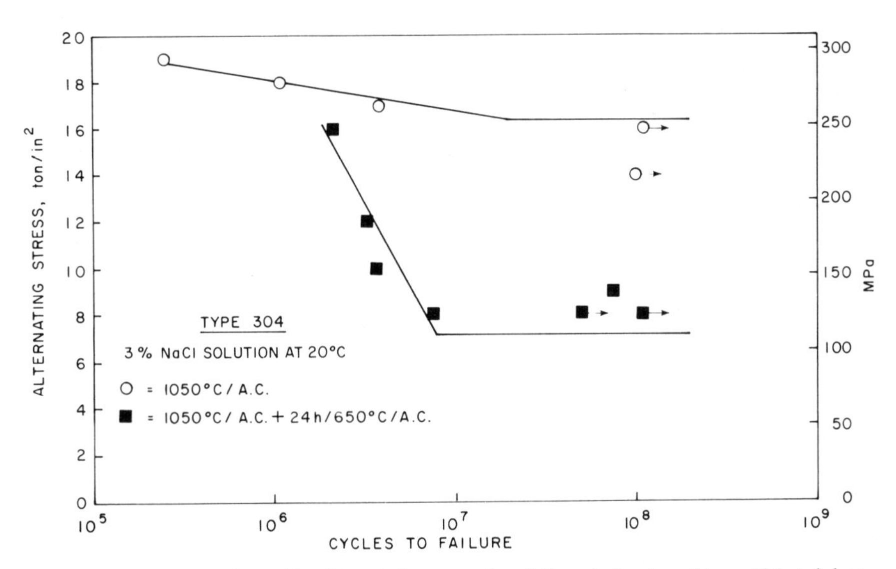

Figure 8-6. Effect of sensitization on the corrosion fatigue behavior of type 304 stainless steel. (Courtesy of Firth-Brown Ltd.)

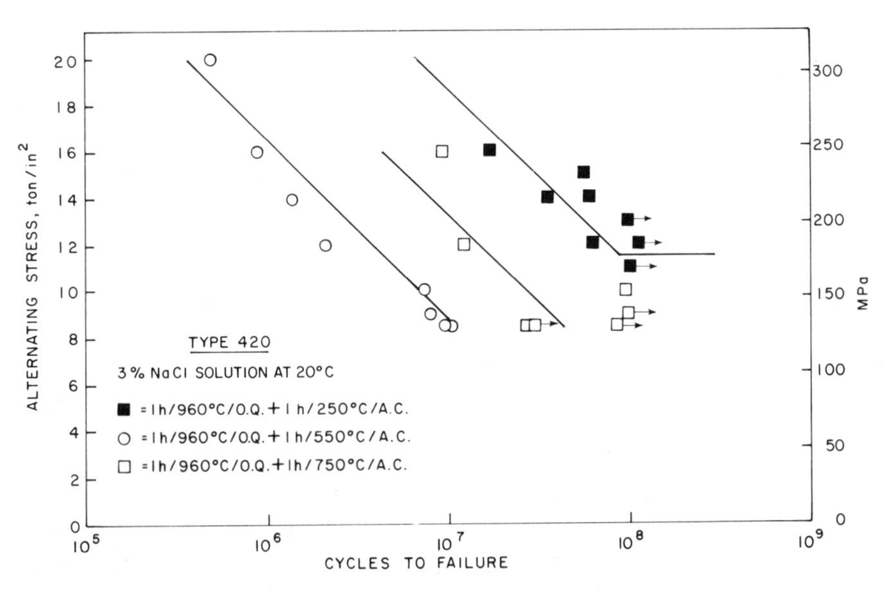

Figure 8-7. Effect of tempering temperature on the corrosion fatigue behavior of type 420 stainless steel. (Courtesy of Firth-Brown Ltd.)

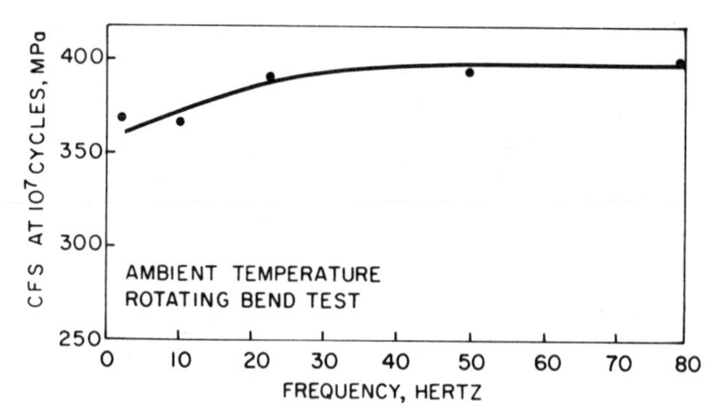

Figure 8-8. Variation of corrosion fatigue strength with frequency in a 0.17 *M* NaCl solution (pH 7.4) of type 316L stainless steel with a small grain size. (After Hughes, Jordan, and Orman.)

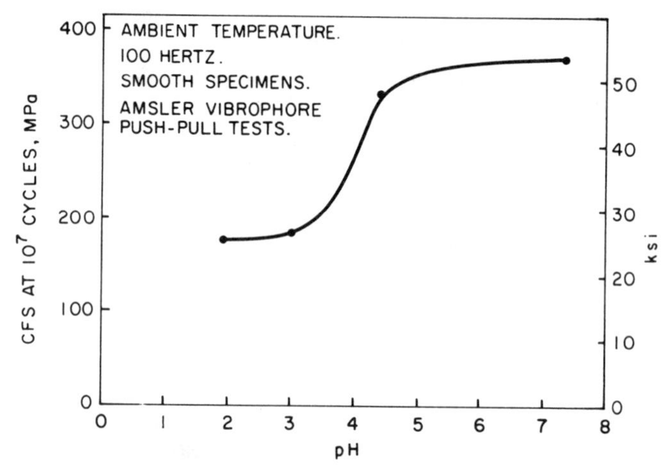

Figure 8-9. Effect of pH of 0.17 *M* NaCl solutions on the corrosion fatigue strength of type 316L stainless steel cold worked 20%. (After Hughes, Jordan, and Orman.)

the CFS at 10^7 cycles for cold worked type 316L in a saline solution. Effects of surface finish on CFS have received some, but not very extensive, study (4).

It should be emphasized that laboratory test data of the type described here should not be used directly to predict the service life of components, such as ships propellers. Both the time of exposure of the component to the corrodent and the magnitude and frequency of the stress cycles may be significantly different for a component in service, and these factors should be taken into consideration.

8-2 GALVANIC CORROSION

When two different metals are immersed in a corrosive solution, each will establish its own corrosion potential, E_{corr}, at which the rates of the anodic reactions will equal the rates of the cathodic reactions (see Chapter 3). If the values of E_{corr} are significantly different for the two metals, and they are in contact or electrically connected, the metal with the more noble E_{corr} will become predominantly cathodic and the metal with the more active E_{corr} will become predominantly anodic. A measurable current will flow between the anode and the cathode. The corrosion rate of the anode will be increased and the corrosion rate of the cathode decreased or entirely stopped. The increased corrosion of the anodic material produced by coupling to a cathodic material is known as galvanic corrosion. The terms bimetallic corrosion or two-metal corrosion are also used. This effect is utilized in the cathodic protection of metals by sacrificial anodes.

A simple practical assessment of whether two different metals are likely to produce galvanic corrosion in a given corrodent can be obtained by measuring and comparing E_{corr} values in that corrodent. If the E_{corr} values differ by hundreds of millivolts, clearly galvanic corrosion is a practical possibility, whereas if the difference is of the order of tens of millivolts galvanic corrosion is less likely. However, the welding of two alloys with small differences in E_{corr} values can lead to increased corrosion of the more active material. While estimating the possibility of galvanic corrosion by comparing E_{corr} values has its drawbacks, since polarization effects are ignored, the errors tend to be on the conservative side.

In an attempt to provide an indication of which metal or alloy combinations are likely to cause galvanic corrosion in seawater, The International Nickel Company has developed a chart of E_{corr} values of various metals and alloys in seawater, which is shown in Figure 8-10. Galvanic coupling of materials exhibiting noble E_{corr} with materials exhibiting active E_{corr} is likely to lead to galvanic corrosion of the active material. Figure 8-10 also illustrates the reason why magnesium, zinc, and aluminum are effective sacrificial anodes for the cathodic protection of steels. Other guidelines, for example, those by Evans and Rance (11), have also been published, based on analysis of data obtained from various sources. The part of the Evans and Rance compilation relating to galvanic attack on stainless steels is reproduced in Table 8-2.

When stainless steels are exposed to fast flowing aerated seawater their E_{corr} values are quite noble, as indicated in Figure 8-10. However, in stagnant seawater deposits may form creating crevices, or a crevice may be formed between the stainless steel and the coupling material. In the

TABLE 8-2. DEGREE OF CORROSION AT BIMETALLIC CONTACTS IN AQUEOUS ENVIRONMENTS

	Contact Metal						
Metal Considered	Gold, Platinum, Rhodium, Silver	Monel(1), Inconel(2), Nickel-Molybdenum Alloys	Cupronickels, Silver Solder, Aluminum-Bronzes, Tin-Bronzes, Gunmetals	Copper, Brasses, "Nickel Silvers"	Nickel	Lead, Tin, and Soft Solders	Steel and Cast Iron
Stainless steels							
Type 304	A	A	A	A	A	A	A
Type 431	C	A or C(s)	A or C(s)	A or C(s)	A	A	A
Type 410	C	C	C	C	B or C	A	A

	Contact Metal								
				Stainless Steels					
Metal Considered	Cadmium	Zinc	Magnesium and Magnesium Alloys (Chromated)	Type 304	Type 431	Type 410	Chromium	Titanium	Aluminum and Aluminum Alloys
Stainless steels									
Type 304	A	A	A	—	A	A	A	A	A
Type 431	A	A	A	A	—	A	A	(o)	A
Type 410	A	A	A	C	C	—	C	C	A

A = The corrosion of the "metal considered" is not increased by the "contact metal."
B = The corrosion of the "metal considered" may be slightly increased by the "contact metal."
C = The corrosion of the "metal considered" may be markedly increased by the "contact metal." (Acceleration is likely to occur only when the metal becomes wet by moisture containing an electrolyte, for example, salt, acid, or combustion products. In ships, acceleration may be expected to occur under in-board conditions, since salinity and condensation are frequently present. Under less severe conditions the acceleration may be slight or negligible.)
s = Serious acceleration of corrosion of type 431 stainless steel in contact with copper or nickel alloys may occur at crevices where the oxygen supply is low.
o = No data available.
1 = Monel alloy 400.
2 = Inconel alloy 600.

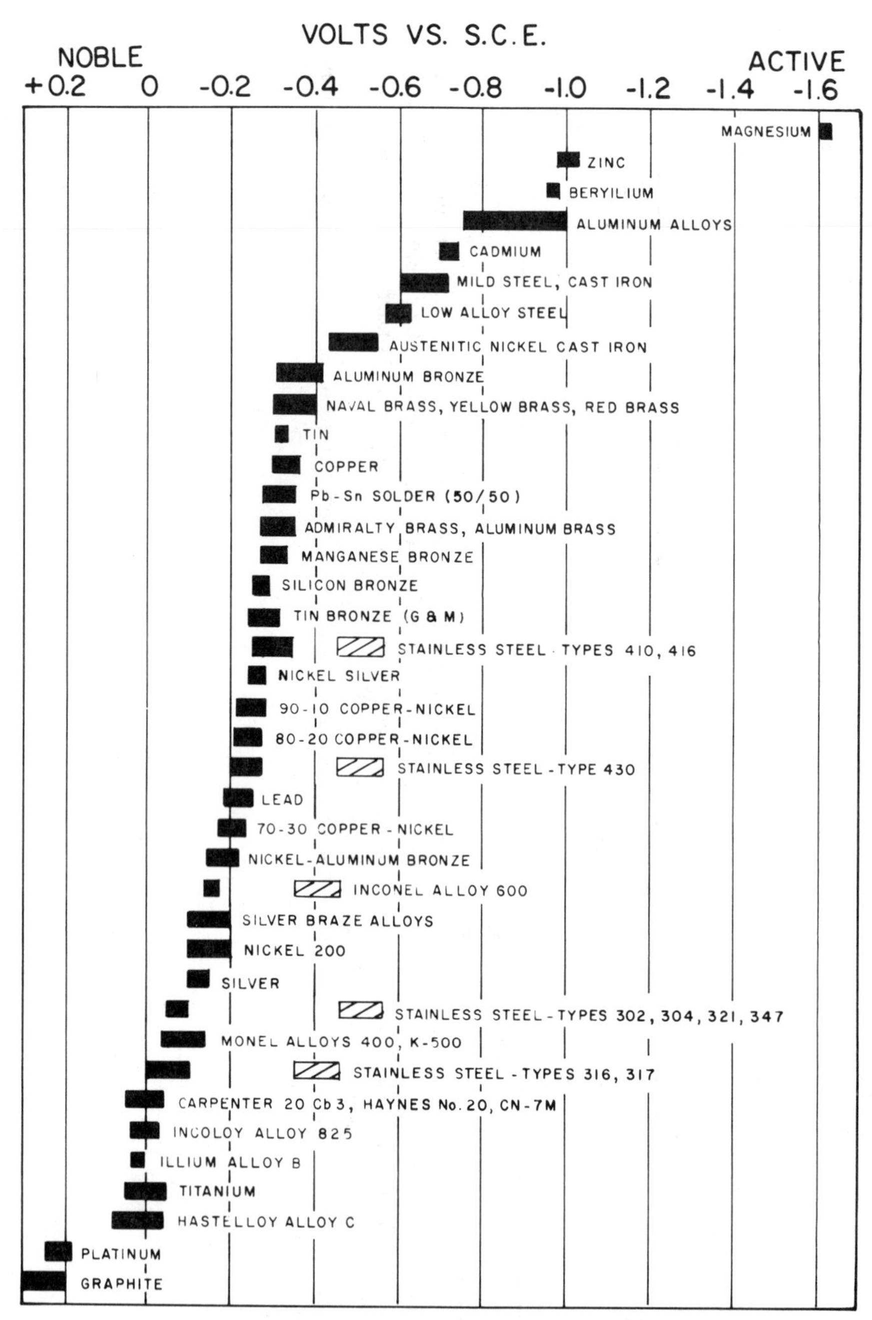

Figure 8-10. Corrosion potentials of various materials in flowing seawater (2.5–4 m/second) at temperatures in the range 10–26°C. The hatched symbols indicate potentials exhibited by stainless steels in acidic water such as exists in crevices.

presence of crevices the stainless steels may exhibit less noble E_{corr} values due to oxygen depletion within the crevice. Under these conditions the less noble E_{corr} values should be used in considering the possibility of occurrence of galvanic corrosion.

A very important factor in determining the extent of attack by galvanic corrosion is the ratio of the areas of the anodic and cathodic materials. An unfavorable area ratio is one which comprises a large cathode and a small anode. Since for a given current flow in a galvanic cell the current density is greater on a small electrode than on a larger one, a small anode will have a greater current density and hence a greater corrosion rate than a large anode. For example, mild steel fasteners (small anodes) should not be used to assemble stainless steel plates (large cathodes).

For listings of materials which are galvanically compatible with stainless steels in fastener applications and in pump and valve trim applications in seawater the reader should refer to the publication by Tuthill and Schillmoller (12), and for a more detailed review of the fundamental understanding of galvanic corrosion, the publication by Pryor (13) should be consulted.

Galvanic corrosion has also been known to cause problems in acid environments. For example stainless steel pickling tanks have been known to corrode when in contact with more active metals being pickled. In terms of the parameters described in Figure 9-1, coupling the stainless steel to a more active metal can move its corrosion potential from $E_{corr\text{-}4}$ (passive) to $E_{corr\text{-}2}$ (active) with a resultant increase in corrosion rate. Hydrogen embrittlement of high strength stainless steel fasteners when in contact with more active metals can be another manifestation of galvanic corrosion.

8-3 EROSION-CORROSION

Erosion-corrosion has been defined by Fontana and Greene (14) as the acceleration of attack caused by a rapidly flowing corrodent sometimes containing solid particles capable of causing erosion or wear. Mildly corrosive environments, even in the presence of solid particles (e.g., seawater containing sand), generally cause little erosion-corrosion on stainless steels. However, in corrosive environments, such as sulfuric acid, the effect of velocity alone can significantly increase corrosion rate (see Chapter 9). Attack in corrosive environments in the presence of solid particles exhibits a highly directional pattern and is found in elbows, bends or T-joints in piping, and propeller blades and pumps.

Since a corrosion effect is superimposed on a mechanical erosion process, resistance to this form of attack is determined by a balance of properties that enhance resistance to corrosion (e.g., alloying with nickel, chromium, and molybdenum) and ones which enhance resistance to erosion (e.g., hardness of the alloy).

Accelerating factors are turbulence (turbulent flow causes greater erosion-corrosion than lamellar flow) and designs that cause impingement (e.g., sharp elbows in piping where the fluid is forced to turn its direction of flow). Obvious remedies include the use of more resistant higher alloys, changes in design, changes in environment, cathodic protection, removal of suspended solids, and reduction in temperature. Weld overlays with hard corrosion resistant materials often provide a practical solution in cases where the other remedial measures are not desirable. For guidelines concerning the avoidance of erosion-corrosion in marine hardware the reader should consult the publication by Tuthill and Schillmoller (12).

8-4 CAVITATION DAMAGE

Cavitation damage is a form of attack that occurs under conditions in which the relative velocity between the liquid and the solid material is very high. In technology it is encountered in pumps, ships propellers, and hydraulic turbines, and a fundamental understanding of the phenomenon requires knowledge of hydromechanics. In essence a liquid experiencing flow divergence, rotation, or vibration forms low pressure areas that generate cavities or bubbles. These cavities form and collapse with extreme rapidity, generating intense shock waves. While the formation and collapse of the liquid cavities is not easy to quantify, Godfrey (15) has noted estimates that two million cavities can collapse within one second over a small area, and that strain waves in material under cavity collapse involve pressures of 1.5 GPa. The collapse of the cavities on a solid surface removes material by mechanical erosion. The observation that cavitation damage can occur on glass and plastics rules out the view that the phenomenon is basically a corrosion process (15). However, corrosion can accelerate cavitation damage of metals and alloys in corrosive environments, and sometimes cause attack under mild cavitating conditions which by themselves only rupture protective surface films.

In considering resistance to cavitation damage in seawater, Tuthill and Schillmoller (12) have noted that the greatest resistance is shown by austenitic stainless steels, precipitation hardening stainless steels, higher alloys, such as Inconel alloy 625, Inconel alloy 718, and Hastelloy alloy C, cobalt base hard facing alloys, and titanium alloys. While all these materials have good corrosion resistance, from a mechanical property viewpoint they range from soft and ductile (e.g., austenitic stainless steels) to hard and high strength (e.g., precipitation hardening stainless steels). In this regard the analysis by Backstrom (16) has shown that a good correlation exists between cavitation resistance and elastic-plastic strain energy (i.e., area under the stress-strain curve obtained in a tensile test), while others claim correlations with functions of hardness.

For severe cavitating conditions redesigning the components to improve the flow characteristics of the liquid may be necessary, since upgrading to more corrosion resistant materials may be of limited use.

REFERENCES

1. D. W. Hoeppner, *Corrosion Fatigue, Chemistry, Mechanics and Microstructure,* National Association of Corrosion Engineers, Houston, Tex., 1972, p. 3.
2. K. Fässler, *l Korrosionum,* Verlag Chemie GmbH, Weinheim/Bergstrasse, Germany, 1974, p. 136.
3. M. O. Speidel, *Handbook on Stress Corrosion Cracking and Corrosion Fatigue,* to be published by Advanced Research Projects Agency, Washington, D.C.
4. H. Spähn, *Corrosion Fatigue, Chemistry, Mechanics and Microstructure,* National Association of Corrosion Engineers, Houston, Tex., 1972, p. 40.
5. J. A. Collins and M. L. Monack, *Mater. Prot. Performance,* Vol. 12, p. 11, 1973.
6. A. J. Sedriks and K. L. Money, *Corrosion Fatigue Properties of Nickel-Containing Materials in Seawater,* The International Nickel Company, Inc., New York, 1978.
7. T. P. May, reported by F. L. LaQue, *Marine Corrosion—Causes and Prevention,* John Wiley & Sons, New York, N.Y., 1975, p. 84.
8. W. W. Kirk, R. A. Covert, and T. P. May, *Met. Eng. Q.,* ASM, p. 31, November 1968.
9. D. B. Franklin, Marshall Space Flight Center Document MSFC-SPEC-522A, November 1977.
10. A. N. Hughes, B. A. Jordan, and S. Orman, *The Corrosion Fatigue Properties of Surgical Implant Materials,* Third Progress Report, AWRE/44/83/140, AWRE, Aldermaston, England, December 1973.
11. U. R. Evans and V. E. Rance, *Corrosion and Its Prevention at Bimetallic Contacts,* brochure published by Her Majesty's Stationery Office, London, 1956.
12. A. H. Tuthill and C. M. Schillmoller, *Guidelines for Selection of Marine Materials,* The International Nickel Company, Inc., New York, 1965.
13. M. J. Pryor, in *Corrosion,* L. L. Shreir, Ed., Vol. 1, Newnes-Butterworths, London, 1976, p. 1.192.
14. M. G. Fontana and N. D. Greene, *Corrosion Engineering,* McGraw-Hill Book Company, New York, 1967, p. 72.
15. D. J. Godfrey, in *Corrosion,* L. L. Shreir, Ed., Vol. 1, Newnes-Butterworths, London, 1976, p. 8.124.
16. T. E. Backstrom, *A Suggested Metallurgical Parameter in Alloy Selection for Cavitation Resistance,* Report No. ChE-72, United States Department of the Interior, Bureau of Reclamation, Denver, Co., December 1967.

9

GENERAL CORROSION

9-1 INTRODUCTION

General corrosion is the term used to describe attack that proceeds in a relatively uniform manner over the entire surface of an alloy. The material becomes thinner as it corrodes until its thickness is reduced to the point at which failure occurs. Stainless steels can exhibit general corrosion in strong acids and alkalies. Failures by general corrosion are less feared because they are often predictable by simple immersion testing or by consulting the corrosion literature. There are numerous publications listing the general corrosion rates of the more commonly used stainless steels in a wide variety of chemical environments (1–6). However, in environments where localized corrosion is predominant, the quoted corrosion rates reflect localized rather than general corrosion.

The development of the mixed potential theory, described in Chapter 3, has provided useful criteria, in terms of readily measurable electrochemical parameters, for determining whether a given stainless steel in a given environment will exhibit general corrosion. Some examples are illustrated in Figure 9-1 in terms of the relative position of the corrosion potential, E_{corr}, with respect to the anodic polarization curve, ABCDE. As discussed in Chapter 3, E_{corr} represents the intersection of the anodic and cathodic polarization curves, and the current density i represents the corrosion rate. The most desirable situation is represented by $E_{corr\text{-}4}/i_4$, which is typical of a stainless steel in water. While some general corrosion is occurring at a rate proportional to i_4, the corrosion rate is very low and the stainless steel is considered to be passive (see Chapter 3). However, low general corrosion rates can be obtained even under conditions where the stainless steel is not passive. For example, the corrosion rate obtained under conditions defined by $E_{corr\text{-}1}/i_1$ is low and may be acceptable for certain industrial operations, such as the containment of relatively dilute

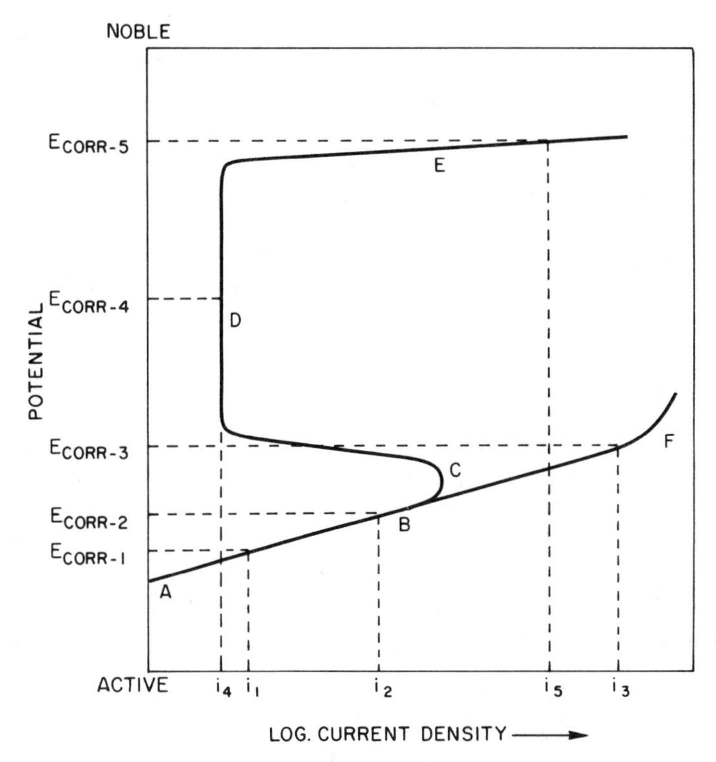

Figure 9-1. Polarization curves showing some examples of general corrosion.

sulfuric acid in stainless steel equipment. More concentrated sulfuric acid will give higher corrosion rates typified by $E_{corr\text{-}2}/i_2$. General corrosion can also occur in the transpassive region, $E_{corr\text{-}5}/i_5$, as would be typified by stainless steel corroding in highly concentrated nitric acid environments. In this regime severe intergranular corrosion may also occur.

It should also be noted that in some environments, such as strong solutions of hydrochloric and hydrofluoric acid, the active-passive behavior characterized by curve ABCDE is not developed, and the current density simply increases with increasing potential, for example, ABF in Figure 9-1. High corrosion rates, typified by $E_{corr\text{-}3}/i_3$, can result under these conditions, particularly in the presence of oxidizers (e.g., Fe^{+3} or Cu^{+2}) which raise E_{corr} in the noble direction. Clearly anodic protection (see Chapter 3) is ruled out under such conditions since raising the potential in the noble direction will simply increase the current density and hence the corrosion rate.

Most cases of general corrosion of stainless steels can be rationalized in terms of the simple criteria described in Figure 9-1, and can be predicted by weight loss techniques or the electrochemical techniques described in Chapter 3. In essence, therefore, the deterioration of stainless steel equipment by general corrosion is in many cases a predictable phenomenon and judgements relating to materials selection for resistance to general corrosion derive primarily from economic considerations, such as the survival of the least costly material for the full design life of given equipment. While failures by localized corrosion often far outweigh failures by general corrosion (e.g., Table 1-3), it is obviously the first step to select materials capable of surviving general corrosion in the bulk environment. A brief guide as to which industrial chemicals are or are not compatible with the common austenitic grades, type 304 and type 316, as well as nickel, is provided by Table 9-1 (3). This listing relates to chemicals under conditions normally encountered in their bulk transportation in marine tankers, barges, railroad tank cars, and highway tank trucks (3). For changes in compatibility resulting from changes in temperature, composition, impurities, or other factors more detailed surveys (1–6) should be consulted. Nevertheless, Table 9-1 serves as a quick guide, particularly with regard to stainless steel/environment combinations that should be avoided.

9-2 EXPRESSIONS OF CORROSION RATE

The rate of general corrosion can either be measured by the electrochemical techniques described in Chapter 3, resistance measurements, or by weight loss tests. In the latter a clean test coupon is measured, weighed, exposed to the corrodent for a known time, cleaned to remove corrosion products, and reweighed. The corrosion rate is calculated from the formula

$$R = \frac{KW}{ATD}$$

where R is the corrosion rate, K is a constant, T is the time of exposure in hours, W is the weight lost in grams, A is the area in square centimeters, and D is the density in grams per cubic centimeter. The corrosion rate can be expressed in a number of different units by using the appropriate value of K. In the U. S. two basically different units are used, one that describes reduction in thickness per unit time, and the other that describes loss in weight per unit area per unit time. Among the former are mils per year, inches per year, inches per month, micrometers per year, and millimeters per year. Among the latter are milligrams per square decimeter per day and grams per square meter per day. The values of K that give these

TABLE 9-1. ALLOY SELECTION GUIDE FOR HANDLING INDUSTRIAL CHEMICALS AND LIQUIDS

Commodity	Nickel	Stainless Steel Type 304	Stainless Steel Type 316
Acetaldehyde	Yes	Yes	Yes
Acetic Acid (Glacial)	No	Yes[d]	Yes[d]
Acetic Anhydride	Yes	Yes	Yes
Acetone	Yes	Yes	Yes
Acetyl Chloride	Yes	No	No
Acrolein	n.a.	Yes	Yes
Acrylic Ester Monomers (methyl and ethyl methacrylate; methyl and ethyl acrylate; butyl methacrylate)	Yes	Yes	Yes
Alcohols (allyl, ethyl, methyl, propyl)	Yes	Yes	Yes
Alum (aluminum sulfate)	No	Yes[d]	Yes[d]
Ammonia, Anhydrous	Yes	Yes	Yes
Ammonia, Aqua	No	Yes	Yes
Ammonium Nitrate Solutions	No	Yes[d]	Yes[d]
Amyl Acetate	Yes	Yes	Yes
Amyl Chloride	Yes	Yes	Yes
Aniline	n.a.	Yes	Yes
Benzaldehyde	Yes	Yes	Yes
Benzene (Benzol Coal Naphtha)	Yes	Yes	Yes
Benzoyl Chloride	Yes	No	No
Benzyl Chloride	Yes	No	No
Bromine	Yes[a]	No	No
Butadiene	Yes	Yes	Yes
Butyl Acetate	Yes	Yes	Yes
Butraldehyde	Yes	Yes	Yes
Butyric Acid	No	Yes[d]	Yes[d]
Caprolactam	No	Yes	Yes
Carbon Bi (or Di) Sulfide	Yes	Yes	Yes
Carbon Tetrachloride	Yes	Yes[b]	Yes[b]
Castor Oil	Yes	Yes	Yes
Caustic Potash (KOH), 50%	Yes	Yes	Yes
Caustic Soda (NaOH), 50%	Yes	Yes	Yes
Caustic Soda (NaOH), 70%	Yes	No	No
Chlorobenzene (Mono or Di)	Yes	n.a.	n.a.
Cocoanut Oil	Yes	Yes	Yes
Corn Syrup	Yes	Yes	Yes
Cresol (Cresylic Acid)	Yes	Yes	Yes
Cumene (Isopropyl Benzene)	Yes	Yes	Yes
Cyclohexanol	Yes	Yes	Yes
Dibutyl Phthalate	Yes	Yes	Yes
Dioctyl Phthalate	Yes	Yes	Yes
Ethyl Alcohol	Yes	Yes	Yes
Ethyl Chloride	Yes	Yes[b]	Yes[b]
Ethyl Ether	Yes	Yes	Yes
Ethylene Dichloride	Yes	Yes[b]	Yes[b]
Ethylene Glycol	Yes	Yes	Yes
Ethylene Oxide	Yes	Yes	Yes
Fatty Acids (Oleic,palmitic,stearic)	Yes	Yes[d]	Yes[d]
Fish Oil	n.a.	Yes	Yes
Fluorine	Yes[a]	No	No
Fluosilicic Acid	No	No	No
Formaldehyde, 37% Sol.	No	Yes	Yes

Table 9-1. (continued)

Commodity	Nickel	Stainless Steel Type 304	Stainless Steel Type 316
Formic Acid	No	Yes[d]	Yes[d]
Fuel Oil	Yes	Yes	Yes
Gasoline	Yes	Yes	Yes
Glucose	Yes	Yes	Yes
Glycerin	Yes	Yes	Yes
Glycolic Acid	No	Yes	Yes
Grape Juice	No	Yes	Yes
Hexane	Yes	Yes	Yes
Hydrazine, Anhydrous	n.a.	Yes	n.a.
Hydrazine, 50% Conc.	n.a.	Yes	n.a.
Hydrobromic Acid, 49% Sol.	No	No	No
Hydrochloric Acid, 38% Sol.	No	No	No
Hydrofluoric Acid, Anhydrous	Yes	Yes	Yes
Hydrofluoric Acid, 60% Sol.	No	n.a.	n.a.
Hydrogen Peroxide	No	No[c]	No[c]
Hydrogen Sulfide	n.a.	Yes	Yes
Isocyanates	Yes	Yes	Yes
Jet Fuels	n.a.	Yes	Yes
Kerosene	Yes	Yes	Yes
Lacquers	Yes	Yes	Yes
Lactic Acid	No	No	No
Latex	No	Yes	Yes
Linseed Oil	Yes	Yes	Yes
Maleic Anhydride	Yes	Yes	Yes
Methyl Acrylate	Yes	Yes	Yes
Methyl Alcohol	Yes	Yes	Yes
Methyl Chloride	Yes	n.a.	n.a.
Methylene Chloride	Yes	Yes[b]	Yes[b]
Methyl Ethyl Ketone	Yes	Yes	Yes
Methyl Methacrylate Monomer	Yes	Yes	Yes
Molasses	Yes	Yes	Yes
Monochlorobenzene	Yes	Yes	Yes
Naphtha	Yes	Yes	Yes
Naphthenic Acids	n.a.	n.a.	n.a.
Nitric Acid, 58-68%	No	Yes[d]	Yes[d]
Nitric Acid, 94-96% (Fuming Grades)	No	No	No
Nitrogen Fertilizer Solutions	No	Yes	Yes
Nitrosyl Chloride	Yes	n.a.	n.a.
Nylon Salt Solution	Yes	Yes	Yes
Oleic Acid	Yes	Yes	Yes
Oleum (See Sulfuric Acid)	No	Yes[d]	Yes[d]
Olive Oil	n.a.	Yes	Yes
Oxygen, Liquid	Yes	Yes	Yes
Palmitic Acid	Yes	Yes	Yes
Palm Oil	n.a.	Yes	Yes
Phenol	Yes	Yes	Yes
Phosphoric Acid, 52-54% P_2O_5	No	No	Yes[d]
Phosphoric Acid, 70-72% P_2O_5 (Superphosphoric)	No	No	Yes[d]
Phosphorus Oxychloride	Yes	Yes[b]	Yes[b]
Phosphorus Trichloride	Yes	Yes[b]	Yes[b]

Table 9-1. (continued)

Commodity	Nickel	Stainless Steel Type 304	Stainless Steel Type 316
Phthalic Anhydride	n.a.	Yes	Yes
Polyethylene Emulsion	Yes	Yes	Yes
Polyvinyl Acetate Water Emulsion	Yes	Yes	Yes
Propylene Dichloride	Yes	Yes	Yes
Propylene Glycol	Yes	Yes	Yes
Propylene Oxide	Yes	Yes	Yes
Seawater	No	No	No
Soap	n.a.	Yes	Yes
Sodium Chlorate, Saturated Sol.	n.a.	Yes	Yes
Sodium Hydrosulfide	Yes	n.a.	Yes
Sodium Lauryl Sulfate	n.a.	n.a.	Yes
Sorbitol	n.a.	Yes	Yes
Soybean Oil	Yes	Yes	Yes
Stearic Acid	Yes	Yes[d]	Yes[d]
Styrene Monomer (Vinyl Benzene)	Yes	Yes	Yes
Sugar (Liquid)	n.a.	Yes	Yes
Sulfur (Molten)	Yes	Yes	Yes
Sulfur Chlorides	Yes	n.a.	n.a.
Sulfur Dioxide	No	Yes	Yes
Sulfuric Acid, 60° Be	No	Yes[d]	Yes[d]
Sulfuric Acid, Oleum	No	Yes[d]	Yes[d]
Sulfuryl Chloride	Yes	n.a.	n.a.
Tall Oil	n.a.	Yes	Yes
Tall Oil Fatty Acids	n.a.	Yes[d]	Yes[d]
Tallow	Yes	Yes	Yes
Thionyl Chloride	Yes	Yes[b]	Yes[b]
Toluene	Yes	Yes	Yes
Trichloroethane	Yes	Yes[b]	Yes[b]
Trichloroethylene	Yes	Yes[b]	Yes[b]
Tricresyl Phosphate	n.a.	Yes	Yes
Tung Oil	n.a.	Yes	Yes
Turpentine (Oil)	Yes	Yes	Yes
UDMH (Unsymmetrical Dimethyl Hydrazine	No	Yes	No
Urea Solution	n.a.	Yes	Yes
Vegetable Oils	n.a.	Yes	Yes
Vinegar	No	Yes	Yes
Vinyl Acetate	Yes	Yes	Yes
Vinyl Chloride Monomer	Yes	Yes	Yes
Waxes	Yes	Yes	Yes
Whiskey	No	Yes	Yes
Wine	No	Yes	Yes
Xylene	Yes	Yes	Yes

CODE:

Yes - Suitable
No - Unsuitable
n.a. - Data not available
a - MONEL alloy 400 is also used.
b - Moisture can cause pitting attack. Steam cleaning can possibly cause stress corrosion cracking.
c - While stainless steel is not corroded by this commodity, long periods of contact might cause deterioration of the commodity.
d. - The extra low carbon type (.03%C max) might be preferred for welded construction.

TABLE 9-2. VALUES OF CONSTANT K TO GIVE VARIOUS CORROSION RATE UNITS

Desired Units	Usual Symbol	Constant K
Mils per Year	mpy	3.45×10^6
Inches per Year	ipy	3.45×10^3
Inches per Month	ipm	2.87×10^2
Millimetres per Year	mm/y	8.76×10^4
Micrometres per Year	µm/y	8.76×10^7
Milligrams per Square Decimetre per Day	mdd	$2.40 \times 10^6 D$
Grams per Square Metre per Day	g/m^2d	$2.40 \times 10^5 D$

D = Density in grams per cubic centimetre.

NOTE: To convert corrosion rate in units, X, to a rate in units, Y, multiply X by K_Y/K_X. For example, for a material with a D = 8, a corrosion rate of 18 mpy is equivalent to:

$$18\ (2.40 \times 10^5 \times 8/3.45 \times 10^6) = 10\ g/m^2d$$

TABLE 9-3. MULTIPLIERS TO CONVERT OTHER UNITS TO MILS PER YEAR

Unit to be Converted	Multiplier
Inches per Year	1,000
Inches per Month	12,000
Millimetres per Year	39.4
Micrometres per Year	0.039
Milligrams per Square Decimetre per Day	1.44/D
Grams per Square Metre per Day	14.4/D

D = Density in grams per cubic centimetre.

NOTE: The following approximate relationship is often useful in cursory comparisons of corrosion rates of stainless steels:

$$10\ g/m^2d = 100\ mdd \approx 20\ mpy \approx 0.5\ mm/y$$

different units when introduced in the corrosion rate equation are listed in Table 9-2. In the U. S. by far the most popular corrosion rate unit has been mils per year (mpy), since it employs English units still in common use, describes reduction in thickness which is a parameter of interest in technology, and describes the corrosion rate of most usable materials on a scale of 1 to 200 without the use of decimals or very large numbers. Multipliers to convert the other units to mils per year are given in Table 9-3. It is likely that the increasing adoption of SI units (Systeme Internationale d'Unites) by U. S. technology will ultimately force the adoption of millimeters per year and micrometers per year to describe corrosion rate in severe and mild corrodents, respectively.

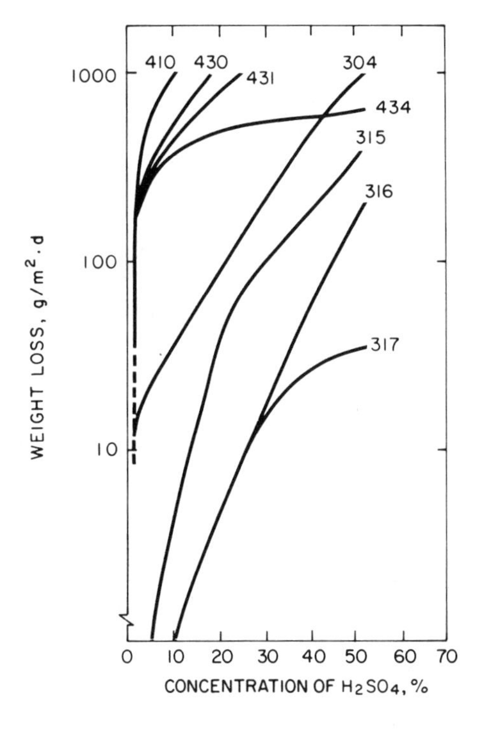

Figure 9-2. Corrosion rates of various stainless steels in different concentrations of undeaerated sulfuric acid at 20°C. (After Truman.)

9-3 ACIDS

9-3-1 Sulfuric Acid

The corrosion rates of a number of stainless steels in different concentrations of undeaerated sulfuric acid at 20°C are shown in Figure 9-2 (7), and the effect of temperature on corrosion rate is given in Figure 9-3 (7). It is evident from these figures that the austenitic grades, particularly the molybdenum grades, have significant resistance at low concentrations and temperatures. The addition of oxidizers such as nitric acid or cupric ions to the sulfuric acid extend the range of usefulness, as indicated in Figure 9-4 (7). In terms of the mixed potential theory the addition of these oxidizers decreases the corrosion rate by changing the corrosion potential/polarization curve relationship from a position defined by $E_{corr\text{-}2}/i_2$ in the absence of oxidizers to a position defined by $E_{corr\text{-}4}/i_4$ in the presence of oxidizers (Figure 9-1). The presence of copper in the alloy, as found in the higher alloys such as Carpenter 20Cb-3 and Incoloy alloy 825 (see Table 2-8) also enhances corrosion resistance in sulfuric acid.

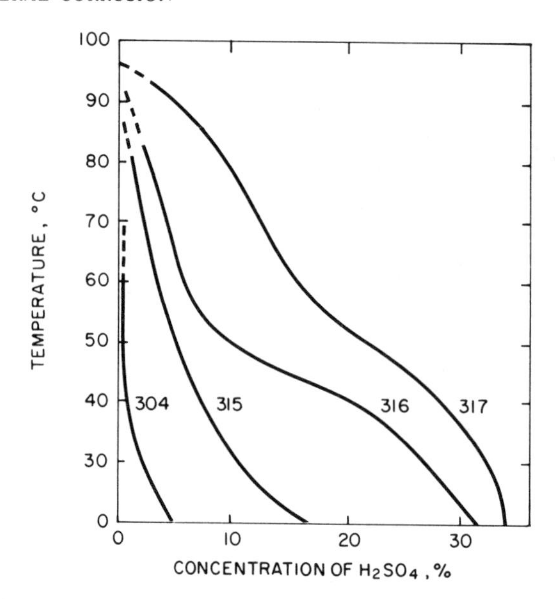

Figure 9-3. Concentrations and temperatures of sulfuric acid solutions required to give a corrosion rate of 25 g/m^2d for various stainless steels. (After Truman.)

Regarding the effect of oxygen in the sulfuric acid, it is generally true to say that austenitic stainless steels in oxygenated solutions of dilute sulfuric acid at ambient temperatures will exhibit passive behavior as defined by the condition $E_{corr\text{-}4}/i_4$ in Figure 9-1. However, there are instances where the introduction of small amounts of oxygen into deaerated acid will increase corrosion rate. An obvious example is illustrated in Figure 9-1. In this case a stainless steel slightly corroding in the active state under conditions typified by $E_{corr\text{-}1}/i_1$ will have its corrosion rate increased by the introduction of small amounts of oxygen if this change produces the situation defined by $E_{corr\text{-}2}/i_2$. However, if the situation defined by $E_{corr\text{-}4}/i_4$ is attained, the corrosion rate will decrease. Because stringent control of the oxygen content is difficult, oxygen is not generally used as a controllable oxidizer for decreasing the corrosion rate. Industrial practice has favored ferric or cupric sulfate, or nitric acid additions when possible.

Increasing the nickel, molybdenum, and copper contents of austenitic alloys results in major increases in corrosion resistance in sulfuric acid environments. A useful perspective of the decreased corrosion rates resulting from alloying with nickel, molybdenum, and copper is provided by the data shown in Figure 9-5 (8). Incoloy alloy 825, Carpenter 20Cb-3, and the cast counterpart CN-7M, are often used in the chemical industry

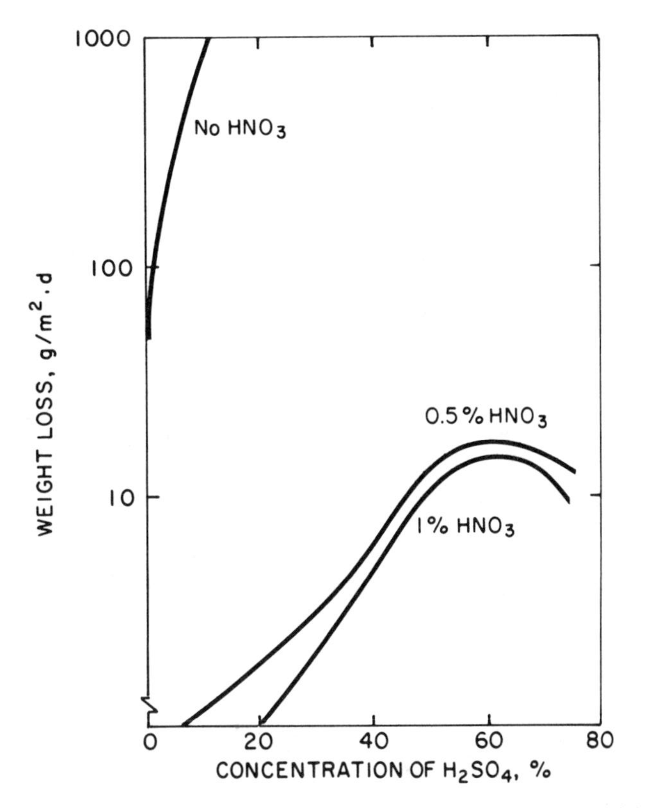

Figure 9-4. Inhibiting effect of nitric acid on the corrosion of type 304 stainless steel in sulfuric acid solutions at 100°C. (After Truman.)

for sulfuric acid service (9). However, numerous other alloys are also used, and for a detailed review the reader should consult reference 10.

It should also be noted that pure sulfuric acid of very high concentrations (e.g., oleum) is not particularly corrosive to austenitic stainless steels (1, 3, 4).

Regarding acid velocity, recent studies by Kain (11), summarized in Table 9-4, have shown that in most cases the corrosion rate increases with increasing velocity. However, there is no simple relationship between the acid velocity and the corrosion rate, with the effects changing with alloy content and temperature. The data of Table 9-4 emphasize that corrosion rates measured under zero velocity conditions cannot be used to predict behavior in flowing acid.

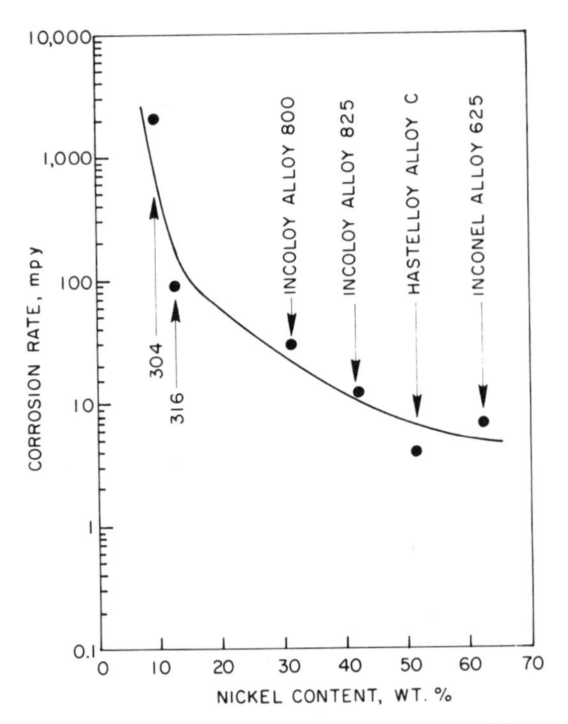

Figure 9-5. Effect of nickel content on the corrosion rate of stainless steels and higher alloys in a 15% H_2SO_4 solution at 80°C. (After Scarberry, Graver, and Stephens.)

TABLE 9-4. EFFECT OF VELOCITY ON CORROSION RATES* OF CAST AND WROUGHT ALLOYS IN 95% SULFURIC ACID AT 50°C AND 70°C, g/m^2d

	Acid Velocity, m/s							
	0		0.6		1.2		1.8	
Alloy	50°C	70°C	50°C	70°C	50°C	70°C	50°C	70°C
Gray Cast Iron	50.9	84.5	115.4	--	135.6	--	124.3	--
1020 Steel	34.3	70.0	193.0	--	285.2	--	367.4	--
CD4MCu	2.9	6.9	2.1	29.1	2.0	38.3	6.6	38.2
CF8	10.4	21.6	10.7	22.8	10.7	29.0	22.3	31.1
Type 304	16.2	20.3	12.3	120.4	11.8	113.0	19.6	83.1
CF8M	10.9	50.7	17.3	86.9	23.6	105.7	40.8	88.3
Type 316	8.9	52.7	17.0	73.2	29.4	87.4	45.8	115.3
CN7M	3.2	5.5	6.7	43.6	7.3	72.9	6.1	77.3
CARPENTER 20Cb-3	5.7	9.2	22.3	59.2	25.2	73.1	15.5	65.4
CW12M-2	0.2	0.6	0.2	1.2	0.1	0.9	0.3	0.7

*Duplicate specimens, average of three 24 h periods.

9-3-2 Hydrochloric Acid

While passivity is possible in very dilute hydrochloric acid solutions, the anodic polarization curve for stainless steels in hydrochloric acid is usually of the type ABF in Figure 9-1. As noted in Section 9-1, under these conditions the introduction of oxidizers increases the corrosion rate. For example, for a stainless steel corroding in hydrochloric acid at position $E_{corr\text{-}2}/i_2$, the addition of oxidizers such as Fe^{+3} or Cu^{+2} will increase the corrosion rate to that typified by $E_{corr\text{-}3}/i_3$.

The extent to which various stainless steels are corroded by various ambient temperature solutions of hydrochloric acid is shown in Figure 9-6 (7), and the effect of temperature is illustrated in Figure 9-7 (7). Figure 9-8 (8) indicates that alloying with nickel and molybdenum to the levels encountered in the higher alloys will increase corrosion resistance. Among Ni-Cr-Fe-Mo alloys (see Table 2-8), Hastelloy alloy C-276 exhibits high resistance to hydrochloric acid environments, as shown in Figure 9-9 (12). Deaeration will decrease and oxygen saturation will increase the corrosion rate of alloy C-276 in hydrochloric acid (12). While a Ni-Mo alloy, known as Hastelloy alloy B, exhibits even higher corrosion resistance in pure hydrochloric acid, its use is not recommended in the pres-

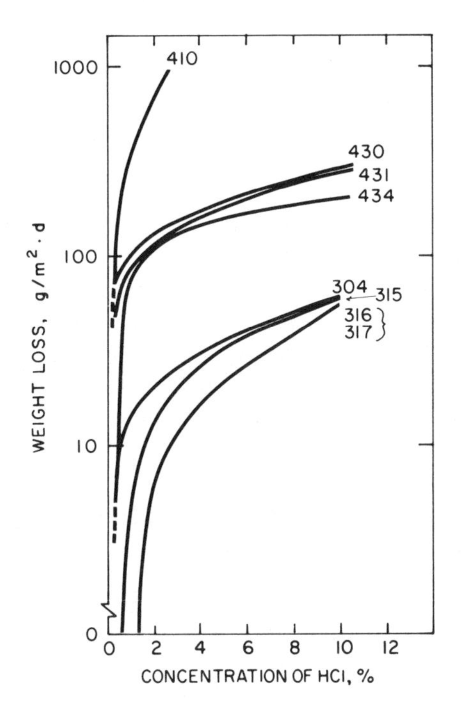

Figure 9-6. Corrosion rates of various stainless steels in hydrochloric acid solutions at 20°C. (After Truman.)

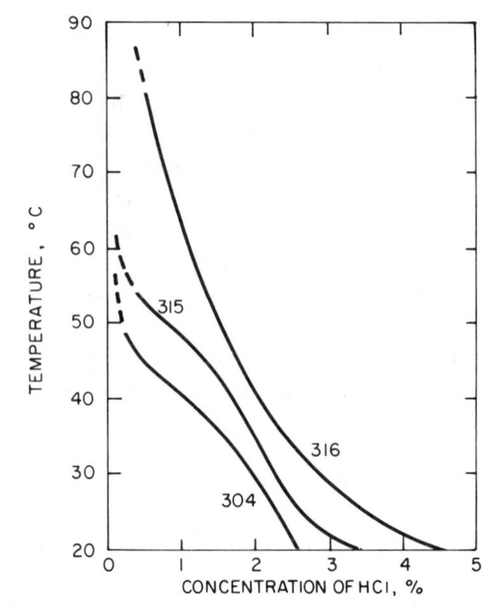

Figure 9-7. Concentrations and temperatures of hydrochloric acid solutions required to give a corrosion rate of 25 g/m²d for various stainless steels. (After Truman.)

ence of Fe^{+3} or Cu^{+2} ions (13). For other alloys used in hydrochloric acid service the reader should consult reference 14.

9-3-3 Phosphoric Acid

Stainless steels in phosphoric acid solutions generally exhibit active-passive behavior of the type characterized by curve ABCDE in Figure 9-1. The presence of oxidizers such as Fe^{+3} or Cu^{+2} often give rise to passive behavior typified by $E_{corr\text{-}4}/i_4$ in Figure 9-1, and the corrosion behavior is determined by the metallurgical and environmental conditions that determine the position of E_{corr}/i. Since in industry phosphoric acid is often encountered in an impure state containing numerous contaminants that can affect E_{corr}/i, testing in process or transportation environments is desirable.

The corrosion rates of various stainless steels in pure boiling solutions of phosphoric acid are shown in Figure 9-10 (7), with types 316 and 317 exhibiting the highest corrosion resistance of the stainless steels tested. Figure 9-11 (8) shows that alloying with nickel and molybdenum gives rise to significant improvements in corrosion resistance. For more detailed discussions of the corrosion resistance of stainless steels and higher alloys

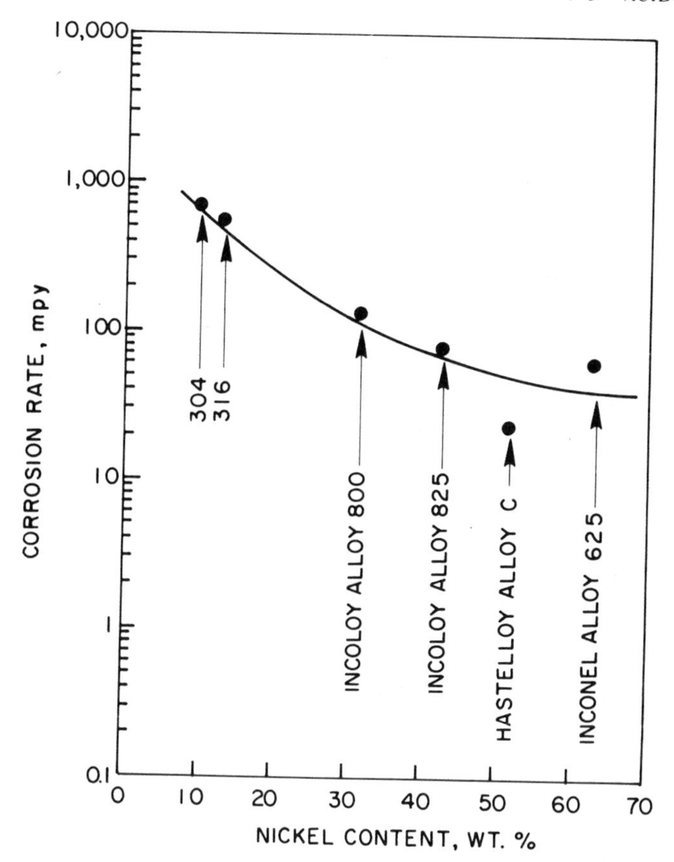

Figure 9-8. Effect of nickel content on the corrosion rate of stainless steels and higher alloys in a 15% HCl solution at 66°C. (After Scarberry, Graver, and Stephens.)

in phosphoric acid environments the reader should consult references 12 and 15–17.

9-3-4 Nitric Acid

In nitric acid stainless steels containing 18% chromium generally exhibit passive behavior (Figure 9-1), $E_{\text{corr-4}}/i_4$, over wide ranges of concentration and temperature. The corrosion rates of types 410, 430, and 304 in boiling nitric acid solutions are shown in Figure 9-12 (7). Problems associated with the intergranular corrosion of sensitized materials (Chapter 6) have led to the widespread use of type 304L for all types of equipment handling nitric acid. The corrosion rates of type 304L as a function of the concentration and temperature of nitric acid are given in Figure 9-13.

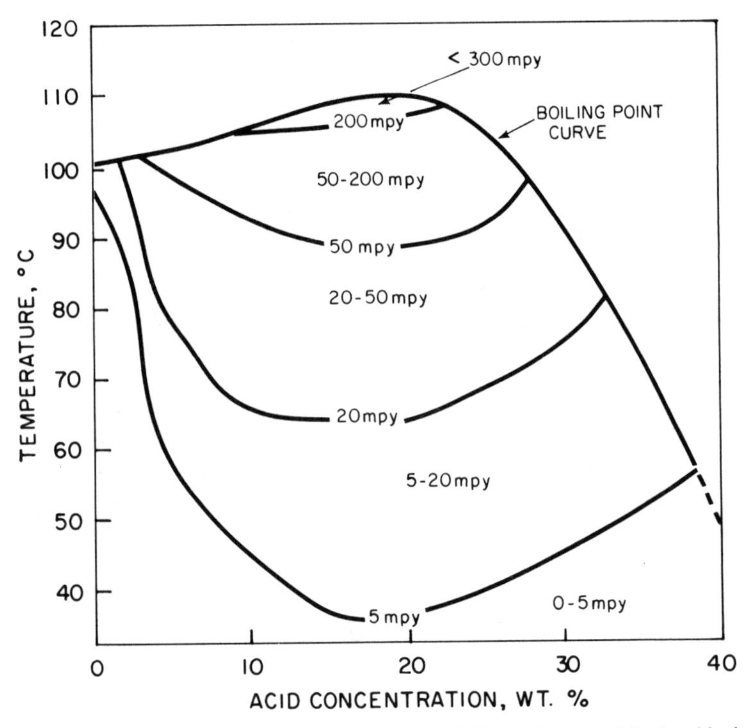

Figure 9-9. Corrosion rates of Hastelloy alloy C-276 in undeaerated hydrochloric acid solutions. (After Lee and Hodge.)

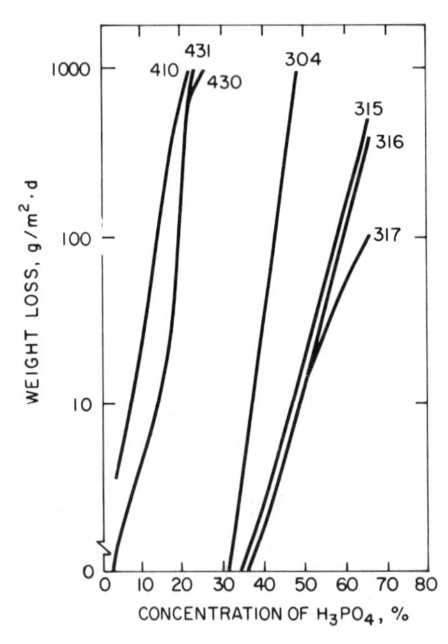

Figure 9-10. Corrosion rates of various stainless steels in phosphoric acid solutions at their boiling points. (After Truman.)

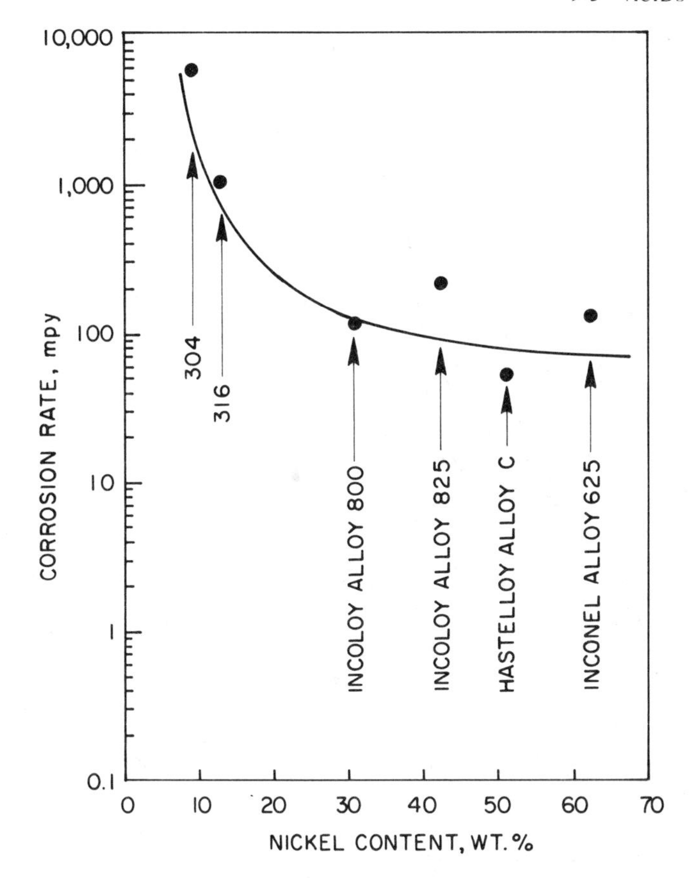

Figure 9-11. Effect of nickel content on the corrosion rate of stainless steels and higher alloys in a boiling 85% H_3PO_4 solution. (After Scarberry, Graver, and Stephens.)

For highly oxidizing conditions in which attack occurs in the transpassive range (Figure 9-1), $E_{corr\text{-}5}/i_5$, or under recirculating conditions under which aggressive corrosion products (e.g., hexavalent chromium ions) can accumulate and cause intergranular attack, high silicon stainless steels may be beneficial. Such high silicon alloys have been described by Coriou et al. (18) and have been used in Europe for concentrated nitric acid service.

Higher chromium alloys (e.g., Inconel alloy 690) are finding increased application for the containment of nitric acid-hydrofluoric acid solutions used for the pickling of stainless steels and reprocessing of nuclear fuels (19).

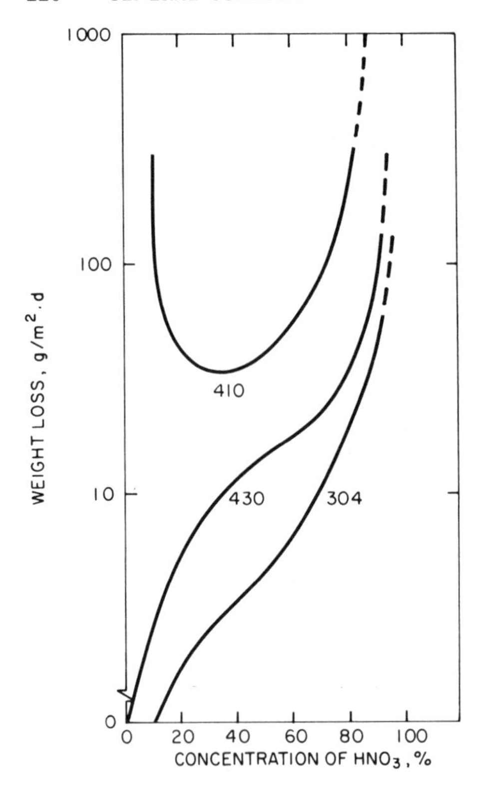

Figure 9-12. Corrosion rates of various stainless steels in boiling nitric acid solutions. (After Truman.)

9-3-5 Organic Acids

Organic acids are generally weaker than inorganic acids because they are less ionized. However, they can be corrosive to stainless steels, and when in the form of aqueous solutions, they can make stainless steels exhibit active-passive behavior. As in the case of sulfuric acid, passive behavior is favored by the presence of oxidizers and active behavior by increasing temperature. Some typical corrosion rates of various stainless steels in acetic and formic acids at various concentrations and temperatures are shown in Table 9-5 (4). However, the corrosion rates in commercial process acids exhibit considerable variation, depending on the contaminants present (20).

Formic acid is one of the more corrosive organic acids, and higher alloys such as Carpenter 20Cb-3 and Incoloy alloy 825 are preferred for handling concentrated hot solutions contaminated with chlorides that may cause pitting.

Acetic acid is less corrosive and can be handled by types 304L and 316L

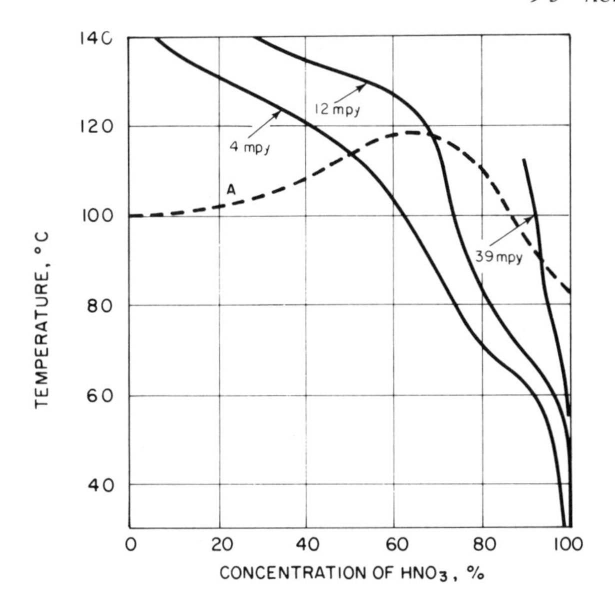

Figure 9-13. Corrosion behavior of type 304L stainless steel in nitric acid solutions of various concentrations and temperatures. Curve A is boiling point curve. (Courtesy of Sandvikens Jernverks Aktiebolag.)

TABLE 9-5. CORROSION OF STAINLESS STEELS BY ORGANIC ACIDS

Acid	Concentration	Temp., °C	Corrosion Rate, mpy				
			410	430	304	316	20Cb-3
Acetic (Air-Free)	50%	24	>50	20-50	<20	< 2	< 2
	50%	100	>50	20-50	20-50	< 2	< 2
	Glacial	24	<20	<20	< 2	20-50	< 2
	Glacial	100	>50	>50	>50	<20	<20
Acetic (Aerated)	50%	24	>50	<20	< 2	< 2	< 2
	50%	100	>50	<20	< 2	< 2	< 2
	Glacial	24	>50	20-50	< 2	< 2	< 2
	Glacial	100	>50	>50	20-50	20-50	<20
Formic	50%	24	20-50	20-50	< 2	>50	< 2
	50%	100	>50	>50	< 2	>50	< 2
	80%	24	>50	20-50	< 2	>50	< 2
	80%	100	>50	>50	< 2	>50	< 2

at low and high temperatures, respectively. The low carbon grades are preferred for fabricated structures to prevent intergranular corrosion. The cast grades CF-8, CF-8M, and CN-7M are used for valves and pumps, with corrosion resistance increasing with molybdenum and nickel content.

For a detailed description of the corrosion resistance of stainless steels

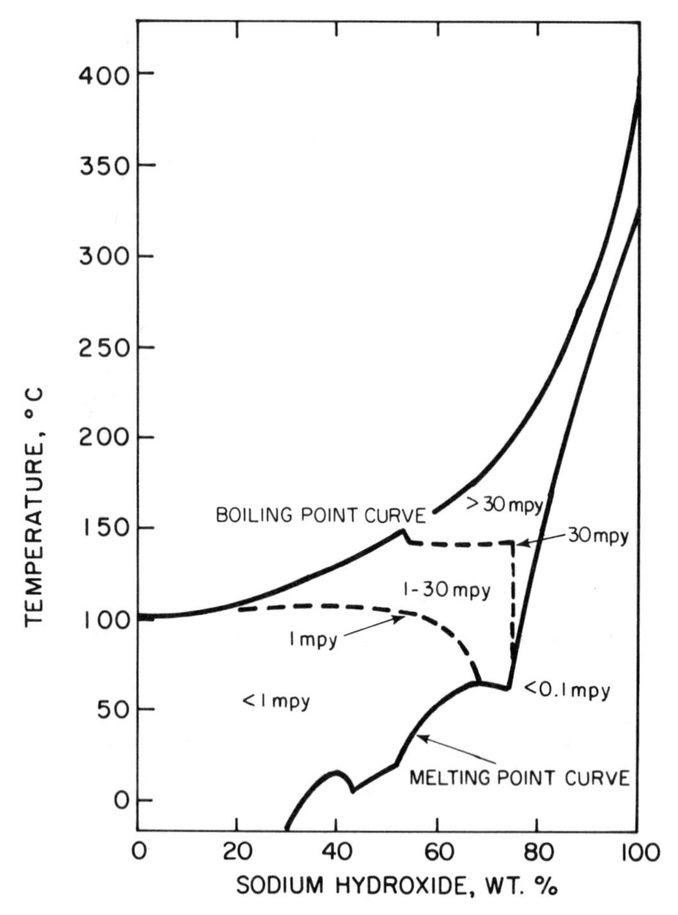

Figure 9-14. Isocorrosion chart for type 304 and 316 stainless steels in sodium hydroxide. (After Swandby.)

and higher alloys in organic acid environments the reader should consult reference 20.

9-3-6 Other Acids

Stainless steels have a rather limited resistance to hydrofluoric acid, the containment of which is generally accomplished by other alloy families (21).

Sulfurous acid, which is generated by the dissolution of sulfur dioxide in water, can often be handled by the molybdenum containing grades of stainless steels (9).

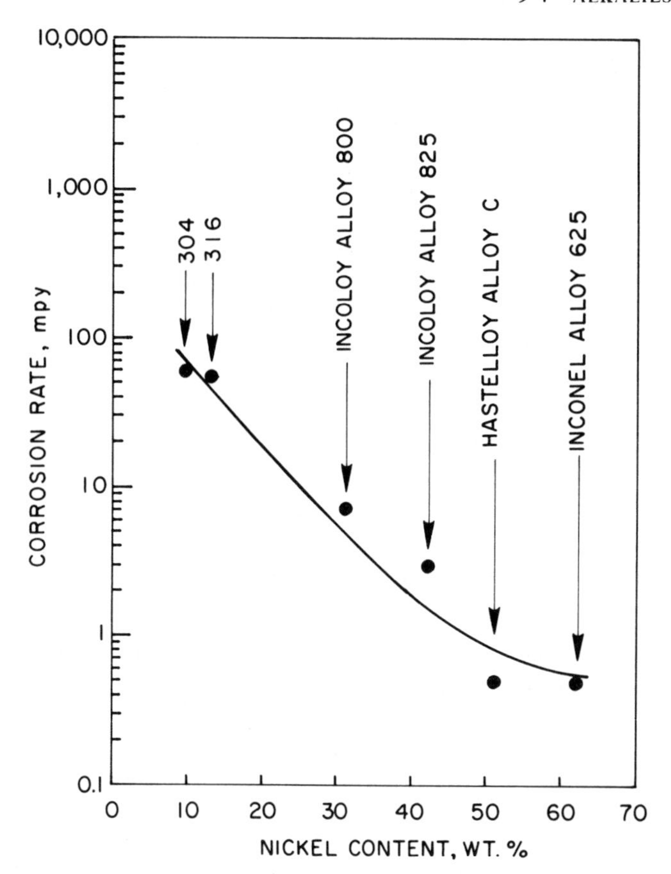

Figure 9-15. Effect of nickel content on the corrosion rate of stainless steels and higher alloys in a boiling 50% NaOH solution. (After Scarberry, Graver, and Stephens.)

Boric acid (22) and carbonic acid (23) generally do not corrode stainless steels.

9-4 ALKALIES

Stainless steels exhibit active-passive behavior in sodium or potassium hydroxide solutions, with the active condition being developed with increasing concentration and temperature. Low general corrosion rates are exhibited by types 304 and 316 over a relatively wide range of temperatures and concentrations, as shown in Figure 9-14 (24). However, at elevated temperatures caustic cracking may occur (see Figure 7-27), and higher nickel alloys or nickel may be preferred. The corrosion rates of

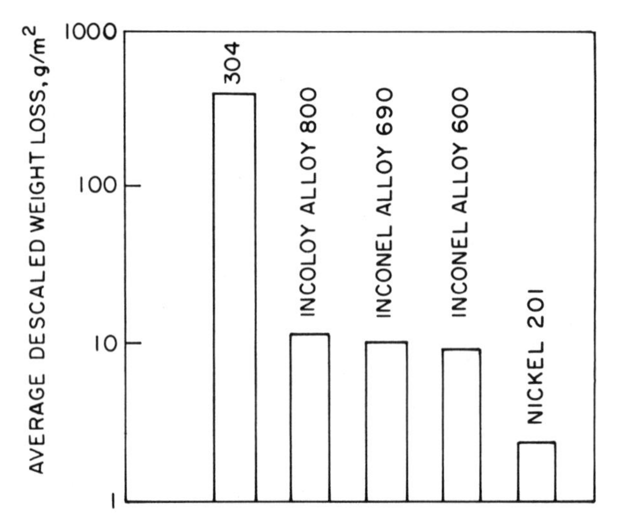

Figure 9-16. Descaled weight loss of various materials after a two week exposure to a deaerated 50% NaOH solution at 316°C. (After Sedriks, Schultz, and Cordovi.)

TABLE 9-6. CORROSION RATES FOR TYPE 304 STAINLESS STEEL AND NICKEL 200 IN BOILING 50% NaOH

		CORROSION RATE DURING EACH 48-HOUR PERIOD (mdd)					
Material	Condition*	Period 1	Period 2	Period 3	Period 4	Period 5	Average
304	MA	970.3	1479.9	1413.0	1361.5	1261.5	1297.3
"	MA	1110.7	1520.7	1495.3	1160.5	1184.4	1294.3
"	MA+L	1128.9	1567.3	1476.4	1315.8	1239.7	1345.6
"	MA+L	1006.3	1510.8	1485.3	1297.7	1272.3	1314.5
Nickel 200**	MA	0.29	1.17	1.61	1.02	2.05	1.23
"	MA	0.88	0.44	1.03	0.44	1.03	0.76

*MA = Mill annealed.
L = 1 hour at 650°C, air cooled.
**Commercially pure nickel, 99.56%Ni-0.04%C.

some of the higher alloys in boiling 50% sodium hydroxide solution are shown in Figure 9-15 (8), and in deaerated 50% sodium hydroxide at 316°C in Figure 9-16 (25). Table 9-6 indicates that commercially pure nickel exhibits very high corrosion resistance in boiling 50% sodium hydroxide. For a discussion of the corrosion resistance of stainless steels and other alloys in caustic environments the reader should consult reference 26.

REFERENCES

1. G. A. Nelson, *Corrosion Data Survey*, The Shell Oil Company, 1960. (Updated and expanded version recently published by NACE, Houston, Tex.)

2. J. P. Polar, *A Guide to Corrosion Resistance,* Climax Molybdenum Company, 1961.
3. *Shipping Today's Chemicals in Modern Metals,* The International Nickel Company, Inc., 1971.
4. *Working Data,* Carpenter Technology Corporation, Reading, Pa., 1973.
5. H. H. Uhlig, *Corrosion Handbook,* John Wiley & Sons, Inc., New York, 1948.
6. E. Rabald, *Corrosion Guide,* Elsevier Publishing Company, New York, 1951.
7. J. E. Truman, in *Corrosion, Metal/Environment Reactions,* L. L. Shreir, Ed., Vol. 1, Newness-Butterworths, Boston, Mass., 1976, p. 3.52.
8. R. C. Scarberry, D. L. Graver, and C. D. Stephens, *Mater. Prot.,* p. 54, June 1967.
9. M. H. Brown, in *Handbook of Stainless Steels,* D. Peckner and I. M. Bernstein, Eds., McGraw-Hill Book Co., New York, 1977, p. 38-1.
10. *Corrosion Resistance of Nickel-Containing Alloys in Sulfuric Acid,* Corrosion Engineering Bulletin CEB-1 (revised), The International Nickel Company, Inc., 1978.
11. R. M. Kain, The International Nickel Company, Inc., to be published.
12. T. S. Lee and F. G. Hodge, *Resistance of Hastelloy Alloys to Corrosion by Inorganic Acids,* Stellite Division, Cabot Corp., Kokomo, Id., 1976.
13. *Corrosion Resistance of Hastelloy Alloys,* Stellite Division, Cabot Corp., Kokomo, Id., 1975.
14. *Resistance of Nickel and High Nickel Alloys to Corrosion by Hydrochloric Acid, Hydrogen Chloride and Chlorine,* Corrosion Engineering Bulletin CEB-3, The International Nickel Company, Inc., New York, 1962.
15. *Corrosion Resistance of Nickel-Containing Alloys in Phosphoric Acid,* Corrosion Engineering Bulletin CEB-4, The International Nickel Company, Inc., New York, 1976.
16. T. Syderberger and S. Nordin, *Corrosivity of Wet-Process Phosphoric Acid,* Paper No. 55, NACE Corrosion/77, San Francisco, March 1977.
17. T. E. Evans, *Stainless Steels for Marine Chemical Tankers—An Investigation of the Influence of Impurities in Wet-Process Phosphoric Acid on the Corrosion Resistance of Types 316L and 317 Stainless Steels,* International Nickel, London, April 1975.
18. H. Coriou, A. Desestret, L. Grall, and J. Hochmann, *Mem. Sci. Rev. Metall.,* Vol. 61, No. 3, 1964.
19. *Inconel Alloy 690,* Huntington Alloys, Inc., Huntington, W. Va., 1973.
20. *The Corrosion Resistance of Nickel-Containing Alloys in Organic Acids and Related Compounds,* Corrosion Engineering Bulletin CEB-6, The International Nickel Company, Inc., New York, 1978.
21. *Corrosion Resistance of Nickel-Containing Alloys in Hydrofluoric Acid, Hydrogen Fluoride and Fluorine,* Corrosion Engineering Bulletin CEB-5, The International Nickel Company, Inc., New York, 1968.
22. G. T. Paul and J. J. Moran, in *Corrosion Resistance of Metals and Alloys,* F. L. LaQue and H. R. Copson, Eds., Reinhold Publishing Corp., New York, 1963, p. 375.
23. *Corrosion Resistance of the Austenitic Chromium-Nickel Stainless Steels in Chemical Environments,* The International Nickel Company, Inc., New York, 1963.
24. R. K. Swandby, *Chem. Eng.,* Vol. 69, p. 186, November 12, 1962.
25. A. J. Sedriks, J. W. Schultz, and M. A. Cordovi, *Corrosion Engineering (Boshoku Gijutsu),* Japan Society for Corrosion Engineering, Vol. 28, p. 82, 1979.
26. *Corrosion Resistance of Nickel and Nickel-Containing Alloys in Caustic Soda and Other Alkalies,* Corrosion Engineering Bulletin CEB-2, The International Nickel Company, Inc., New York, 1976.

10

CORROSION BY HOT GASES AND MOLTEN COMPOUNDS

10-1 INTRODUCTION

Although stainless steels are popular materials for high temperature service, they will react with most high temperature gases to form surface scales. Their usefulness in any given application will be governed by the rate at which the gas/metal reaction can take place through the scale. Conditions that give rise to very slow scale growth are obviously desirable. Conditions that favor rapid scale growth are undesirable since rapid scale formation will quickly consume the stainless steel until the load bearing section is reduced to the point at which mechanical failure occurs.

In some gases (e.g., halogens) volatile reaction products can be formed which vaporize from the surface, while in other cases there is preferential brittle oxide formation along grain boundaries. Accordingly, it is not possible to describe attack by high temperature gases in terms of any single unifying mechanism. However, it is possible to identify various forms of attack in terms of specific gas/metal reactions. Among those of most concern in technological applications are oxidation, sulfidation, carburization, nitriding, and attack by halogen gases. It should also be recognized that in the complex process streams encountered in modern technology multiple modes of attack may occur, and that there is an increasing need to develop theoretical treatments capable of handling multicomponent gaseous systems.

Fuel ash corrosion is a form of attack caused by molten vanadium and sodium compounds. These melt at relatively low temperatures and destroy the protective oxide scales.

10-2 OXIDATION

10-2-1 Introduction

Oxidation, which in the terminology used in gas/metal reactions refers to the formation of oxide scales, is the most frequent cause of high temperature corrosion of stainless steels. It can occur in oxygen, air, carbon dioxide, or steam, or in more complex industrial atmospheres containing significant quantities of these gases. The growth of the scales (i.e., corrosion rate) is determined by numerous metallurgical and environmental parameters and the properties of the scales themselves.*

There are several ways of measuring the extent of oxidation. Measurement of weight change per unit area in a given time has been a popular procedure. It should be noted that, unless the scale spalls off, oxidation is accompanied by weight gain. Because some of the scale can be lost by spalling, some researchers prefer to remove the scale by mechanical means and report attack as "descaled weight loss." Metallographic evaluations of the extent of attack have several advantages over weight change techniques. A metallographic technique used in our laboratory to determine the extent of oxidation (and as discussed later, also sulfidation) measures parameters termed "metal loss" and "maximum attack," as shown in Figure 10-1. This type of data is particularly valuable since it provides information about the reduction of the load bearing section (metal loss) and provides some insight into the extent of grain boundary oxidation which can also affect structural integrity. The parameters shown in Figure 10-1 relate to cylindrical specimens. In sheet specimens of aluminum containing ferritic stainless steels the extent of edge attack should also be evaluated since geometric factors at edges can lead to localized scale breakdown.

10-2-2 Characteristics of Oxidation

High resistance to oxidation of stainless steels is generally associated with the formation of chromic oxide, Cr_2O_3. This oxide is usually not pure chromic oxide and may also contain small amounts of iron and nickel. It is formed on most stainless steels under mildly oxidizing conditions and on high chromium stainless steels (e.g., type 310) under more severe oxidizing conditions. This oxide grows at a very slow rate, since cation diffu-

*For more detailed treatments of gas/metal reactions the reader is referred to books by Kofstad, "High Temperature Oxidation of Metals" (John Wiley & Sons), Haufe, "Oxidation of Metals" (Plenum Press), Kubaschewski and Hopkins, "Oxidation of Metals and Alloys" (Butterworth & Co.), and Mrowec and Werber, "Gas Corrosion of Metals" (U.S. National Technical Information Service, TT76-54038).

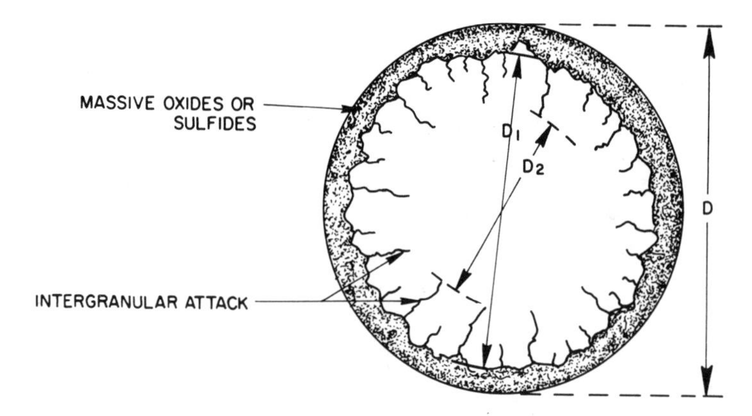

D = ORIGINAL DIAMETER, MEASURED WITH A MICROMETER.

D_1 = DIAMETER OF STRUCTURALLY USEFUL METAL. MEASURED AT 100X.

D_2 = DIAMETER OF METAL UNAFFECTED BY OXIDES AND SULFIDES, MEASURED AT 100X.

METAL LOSS: $D-D_1$ LOSS IN DIAMETER DUE TO MASSIVE OXIDES AND SULFIDES.

MAXIMUM ATTACK: $D-D_2$ LOSS IN DIAMETER DUE TO ALL FORMS OF OXIDATION AND SULFIDATION.

Figure 10-1. Method of measuring hot corrosion attack.

sion through it, which is thought to be the process controlling its growth, is very slow. Lower chromium stainless steels (e.g., type 304) may form the spinel oxide $FeCr_2O_4$ (sometimes described by the equation $FeFe_{2-x}Cr_xO_4$ with x approaching the value of 2), and under certain conditions this spinel oxide may also be relatively protective. However, it is probably true to say that the formation and maintenance of Cr_2O_3 provides the most protective situation.

The initial presence of Cr_2O_3 scale does not guarantee high oxidation resistance as a function of time. As noted by Wood (1) and as shown Figure 10-2, a number of possible changes can take place in the oxidation behavior with time. Under mild oxidizing conditions the protective Cr_2O_3 scale is maintained, with its growth kinetics approximating a parabolic relationship, curve OAD. Yet under more severely oxidizing conditions oxide growth may initially follow curve OA, but at some point a sudden increase in growth rate may occur, as indicated by curve AB. This is known as a "breakthrough," and its occurrence is dependent on alloy composition, environment, and time. As noted by Morris (2), this breakthrough corresponds with the formation of a duplex scale consisting of an inner layer of a spinel oxide ($FeCr_2O_4$) and an outer layer of ferric

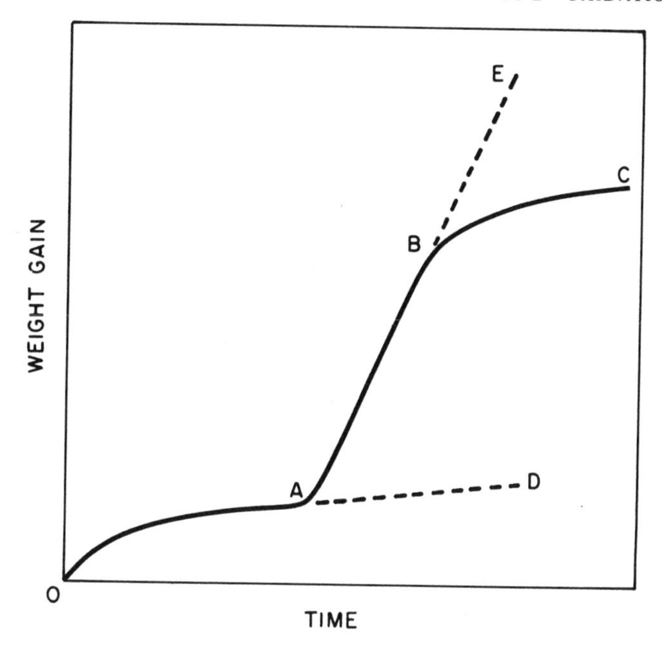

Figure 10-2. Typical oxide growth curves for stainless steels. (After Wood.)

oxide (Fe_2O_3). For discussions of the mechanism of this breakthrough the reader should consult references 3–8.

Depending on alloy composition and environmental conditions, the rapid oxidation may continue, as indicated by curve ABE, or the oxidation rate may again decrease, as shown by curve BC. As the oxide thickens along the curve BC, cracking of the scale may give rise to further surges of rapid oxidation, particularly when the scale gets thicker than 10 μm (6). At very high temperatures (e.g., 1200°C) the Cr_2O_3 may begin to volatilize.

Clearly from a practical viewpoint the most desirable situation is that defined by curve OAD, namely, the formation and maintenance of the very slowly growing Cr_2O_3 scale on stainless steels.

10-2-3 Effects of Composition

The effects of various alloying elements on the oxidation resistance of stainless steels are summarized in Figure 10-3, with the data being taken mostly from the review paper by Morris (2). In this figure the elements are described as having either a beneficial, detrimental, or a variable effect. It is also evident that relatively few elements have been studied.

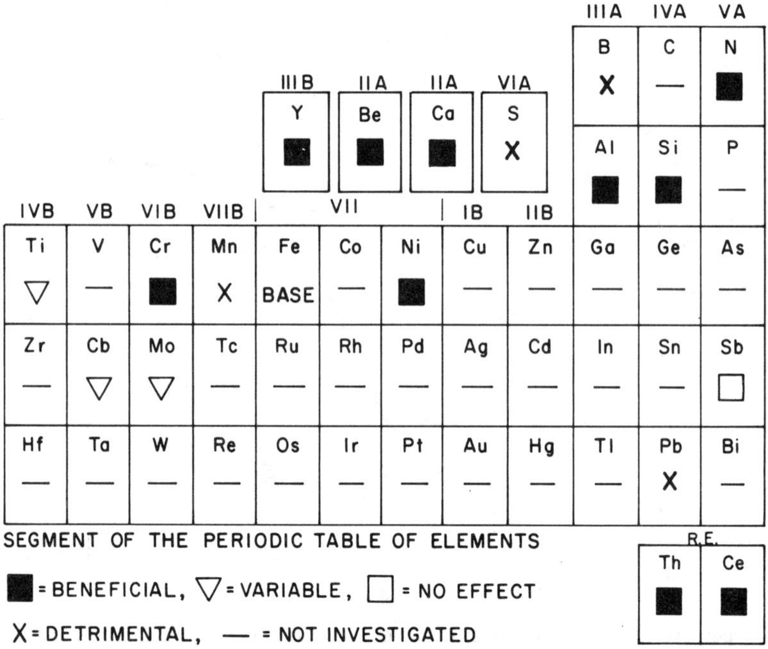

IIIB	IIA	IIA	VIA	IIIA	IVA	VA
				B X	C —	N ■
Y ■	Be ■	Ca ■	S X	Al ■	Si ■	P —

IVB	VB	VIB	VIIB	VII	VII	VII	IB	IIB	IIIA	IVA	VA
Ti ▽	V —	Cr ■	Mn X	Fe BASE	Co —	Ni ■	Cu —	Zn —	Ga —	Ge —	As —
Zr —	Cb ▽	Mo ▽	Tc —	Ru —	Rh —	Pd —	Ag —	Cd —	In —	Sn —	Sb □
Hf —	Ta —	W —	Re —	Os —	Ir —	Pt —	Au —	Hg —	Tl —	Pb X	Bi —

R.E.	
Th ■	Ce ■

SEGMENT OF THE PERIODIC TABLE OF ELEMENTS

■ = BENEFICIAL, ▽ = VARIABLE, □ = NO EFFECT

X = DETRIMENTAL, — = NOT INVESTIGATED

Figure 10-3. Effect of element shown on resistance of stainless steels to oxidation.

In stainless steels, chromium is by far the most important element for enhancing oxidation resistance. The formation of the protective Cr_2O_3 scale involves the selective oxidation of chromium at the metal surface, resulting in the depletion of chromium at the metal/oxide interface. To maintain or stabilize the Cr_2O_3 scale it is important that the bulk chromium content be sufficiently high so that it does not fall below certain minimum values in this interface region. These minimum values for various laboratory Fe-Cr-Ni and Fe-Cr alloys have been determined by Croll and Wallwork (9) and vary with alloy composition. For commercial stainless steels bulk chromium contents approaching those of type 310 (i.e., 25% Cr) are needed to ensure good oxidation resistance at temperatures of the order of 1000°C under cyclic temperature exposures, as indicated in Figure 10-4 (10).

Comparing the oxidation resistance of stainless steels containing the same amounts of chromium and silicon but different amounts of nickel, for example, as given in Table 10-1 (2), shows a pronounced beneficial effect of nickel on oxidation resistance. A number of explanations have been put forward to account for this. It has been suggested that nickel

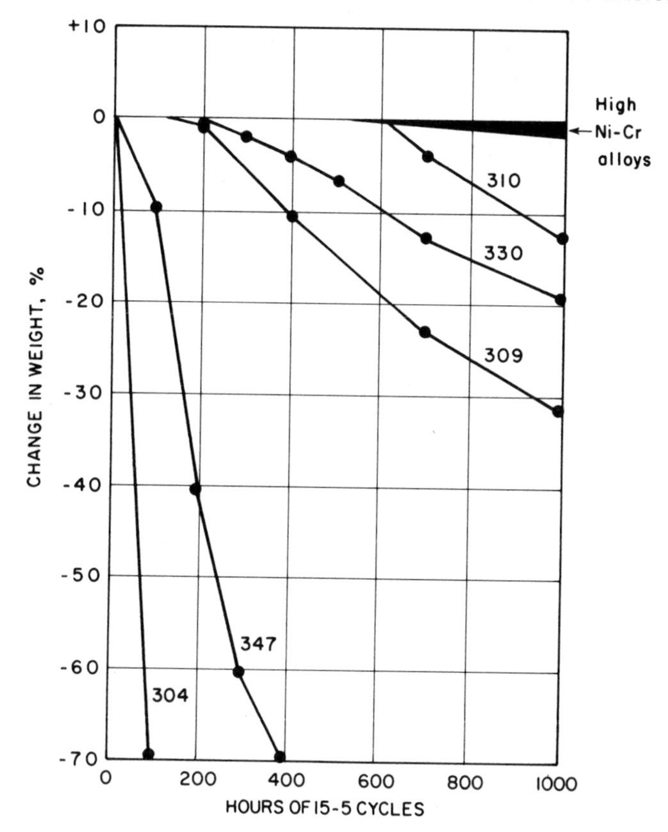

Figure 10-4. Scaling resistance of some Fe-Cr-Ni alloys in air at 980°C. Test employed cyclic exposures with each cycle consisting of 15 minutes at 980°C and 5 minutes cooling. (After Eiselstein and Skinner.)

influences the adhesion and mechanical properties of the scale (11), reduces the rate of cation diffusion in Cr_2O_3 scales (1), and retards the breakthrough transformation, Cr_2O_3 to $FeCr_2O_4 + Fe_2O_3$ (12).

The effect of silicon is also designated as beneficial in Figure 10-3. While there is some question as to whether small amounts of silicon (0.5%) improve the oxidation resistance of laboratory prepared Fe-26% Cr ferritic alloys (13), there is good evidence that silicon improves the oxidation resistance of commercial austenitic stainless steels, as shown in Figure 10-5 (14). The increased oxidation resistance of type 302B (3% Si maximum) over type 302 (1% Si maximum) is readily in evidence, although it falls short of the oxidation resistance exhibited by the high chromium type 310. It has been suggested that the beneficial effect of

TABLE 10-1. EFFECT OF NICKEL CONTENT ON THE GENERALLY ACCEPTED MAXIMUM SERVICE TEMPERATURE IN AIR FOR TWO CAST STAINLESS STEELS

Stainless Steel	Composition, Wt. % Ni	Cr	Si	Maximum Continuous Service Temp., °C
HF	10.5	21	2	900
HN	25	21	2	1095

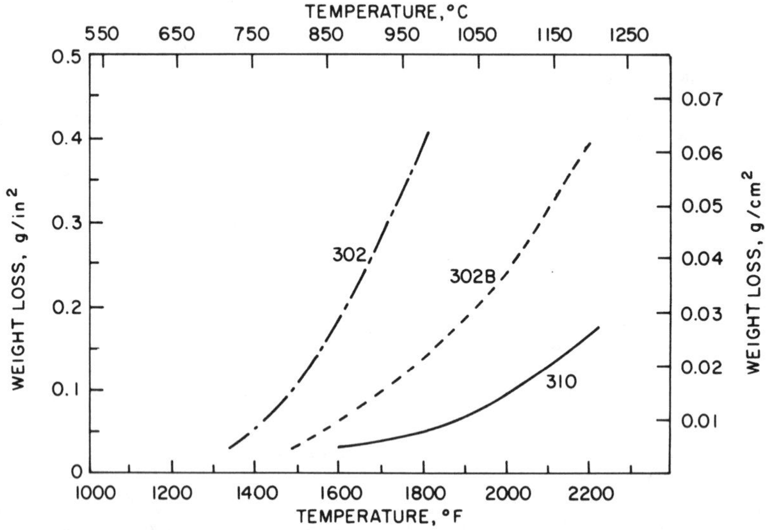

Figure 10-5. Scaling losses developed in 12 intermittent heating and cooling cycles by various stainless steels. (After Grodner.)

silicon is found only in the presence of manganese (15), and the latter element is present in commercial stainless steels. While the mechanism by which silicon improves oxidation resistance remains to be established, it has been suggested that silicon aids the development of a chromium rich scale (16).

It is well established that aluminum additions above certain critical amounts improve the oxidation resistance of wrought and cast stainless steels under conditions of continuous (isothermal) exposure (17), although performance under cyclic exposures has been questioned (18). Kvernes, Oliveira, and Kofstad (19) have recently examined the oxidation behavior of Fe-13% Cr-*x*% Al alloys in air-water vapor mixtures. They have found that at certain critical aluminum concentrations protective scales of α-Al_2O_3 are formed, with the critical aluminum concentration increasing with increasing temperature. Thus, at 980°C, 4% Al is necessary, while at 680°C, 1% Al provides excellent oxidation resistance. They also found that when continuous Al_2O_3 scales are formed, oxidation resistance is not significantly dependent on water vapor content.

It should also be pointed out that aluminum promotes the formation of delta ferrite (see Chapter 1) and the brittle sigma phase, and may reduce cold formability. Therefore the use of aluminum additions to improve oxidation resistance should be carefully weighed against other metallurgical requirements. Because of this, proprietary stainless steels and higher alloys that employ aluminum to enhance oxidation resistance either contain relatively low chromium levels [e.g., MF-2 (12% Cr), OR-1 (13% Cr), and 18SR (18% Cr)] or high nickel levels [e.g., Inconel alloy 601 (59% Ni)]. For a comparison of the oxidation resistance of these materials the reader should consult the publication by Michels (20).

Laboratory studies have shown that the additions of small amounts of cerium, yttrium, and thorium improve the cyclic oxidation resistance of stainless steels (18, 21, 22). A number of theories have been put forward to explain this effect, including the keying of the external scale by the cerium, yttrium, and thorium subscales (2) and the provision of vacancy sinks by the subscale particles which prevent the formation of voids at the oxide/metal interface (23).

Calcium has also been shown to be beneficial to oxidation resistance, as illustrated in Figure 10-6 (24), with the benefit increasing with decreasing chromium content. Mrowec and Werber have reported USSR studies which have identified beryllium and nitrogen to be beneficial, sulfur, lead, and boron to be detrimental, and antimony to have no effect.*

Manganese has been reported to be detrimental to the oxidation resistance of stainless steels (13, 16), with a spinel oxide, $MnO \cdot Cr_2O_3$ being formed in preference to the protective Cr_2O_3 scale.

Titanium, columbium, and molybdenum are designated as having a variable effect on the oxidation resistance of stainless steels in Figure 10-3. Although additions of titanium and columbium of the order of 2% reduce the oxidation resistance of cast stainless steels (15), their presence at levels found in the stabilized grades 321 and 347 does not require the lowering of the generally accepted maximum service temperatures in air below that used for type 304, as shown in Table 10-2 (2).† The same

*Mrowec and Werber, *Gas Corrosion of Metals*, U.S. National Technical Information Service, TT76-54038, 1978.

†Based on a rate of oxidation of 10 mg/cm^2 in 1000 hours, or on service experience.

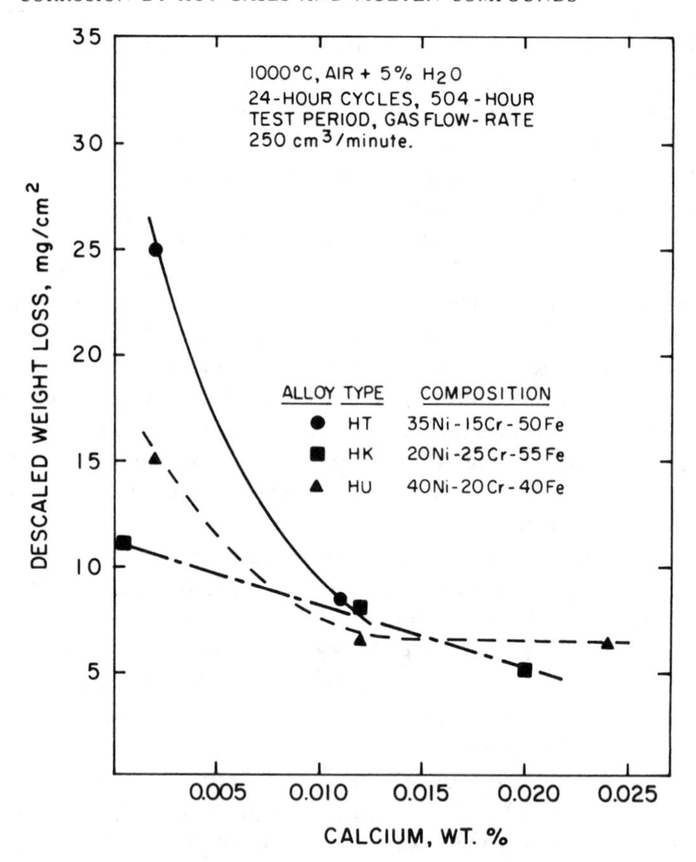

Figure 10-6. Effect of calcium content on the cyclic oxidation resistance of selected heat resistant alloys. (After deBarbadillo.)

applies for molybdenum at the levels found in types 316 and 317 (Table 10-2), although large molybdenum additions (i.e., in excess of 5%) lead to severe oxidation (17, 25) because of the formation of volatile molybdenum oxides.

10-2-4 Effects of Microstructure

The generally accepted maximum service temperatures in air for wrought austenitic, martensitic, and ferritic stainless steels are shown in Table 10-2 (2) and for cast stainless steels in Table 10-3 (2). Since oxidation resistance is dominated by composition rather than the lattice structure of the alloy, the rankings primarily reflect the effects of chromium and nickel contents. These temperatures are identified as either for continuous ser-

TABLE 10-2. GENERALLY ACCEPTED MAXIMUM SERVICE TEMPERATURES IN AIR FOR WROUGHT STAINLESS STEELS

Material	Intermittent Service °C	Continuous Service °C
	AUSTENITIC	
201	815	845
202	815	845
301	840	900
302	870	925
304	870	925
308	925	980
309	980	1095
310	1035	1150
316	870	925
317	870	925
321	870	925
330	1035	1150
347	870	925
	MARTENSITIC	
406	815	1035
410	815	705
416	760	675
420	735	620
440	815	760
	FERRITIC	
405	815	705
430	870	815
442	1035	980
446	1175	1095

TABLE 10-3. GENERALLY ACCEPTED MAXIMUM SERVICE TEMPERATURES IN AIR FOR CAST STAINLESS STEELS

Material	Wrought Comparative	Compositions Preferred for Cyclic Service	Continuous Service °C
HA	--		650
HC	446	Good	1120
HD	--		1065
HE	--		1065
HF	302B		900
HH	309		1065
HI	--		1120
HK	310		1095
HL	--		1150
HN	--	Good	1095
HT	330	Very Good	1035
HU	--	Very Good	1095
HW	--	Excellent	1095
HX	--	Excellent	1150

vice or intermittent service. Intermittent service involves thermal cycling which can cause the scales to crack and spall because of differences in the expansion and contraction characteristics between the metal and the scale. Accordingly, the continuous service temperatures would be expected to be higher than the intermittent service temperatures, and this is indeed the case for the austenitic stainless steels, as indicated in Table 10-2. However, for the ferritic and most of the martensitic stainless steels the generally accepted maximum service temperatures are higher for intermittent service than for continuous service (Table 10-2), implying that for these materials more protective scales are formed under cyclic conditions. Further studies to explain this behavior would seem desirable.

As noted in the Section 10-1, grain boundaries can provide sites for accelerated oxidation. Studies by Keith, Siebert, and Sinnott (26) of the grain boundary oxidation of alloys having a composition near type 310 found no difference between the composition of surface scales and the grain boundary oxides, but found boundary oxidation to increase with temperature. These authors concluded that since the difference between volume and boundary diffusion coefficients decreases with increasing temperature, grain boundary oxidation cannot be a process controlled by boundary diffusion, and they argued that compositional or structural differences must account for the preferential oxidation at grain boundaries. However, Colombier and Hochmann (27) have noted that decreasing the grain size of Ni-Cr alloys increases their oxidation resistance, and this observation could be explained in terms of increased rate of chromium diffusion to the surface along boundary paths. The role of grain boundaries in the oxidation of stainless steels appears to be a subject deserving further study.

Cold work is reported to improve the oxidation resistance in the presence of water vapor of both ferritic (28) and austenitic (29) stainless steels. However, from a mechanistic viewpoint, it is not clear whether this improvement is related to the presence of the deformed microstructure or surface effects, or both.

In considering the subject of microstructure for high temperature applications in oxidizing environments, good oxidation resistance is obviously of great importance. However, other important variables are the intrinsic high temperature mechanical properties of the materials. It would be futile to select a stainless steel with high oxidation resistance for an application in which strength requirements could not be met. In general austenitic stainless steels are substantially stronger than ferritic stainless steels at high temperatures, as indicated by a comparison of stress rupture properties, shown in Figure 10-7, and creep properties, illustrated in Figure 10-8 (30). Embrittlement due to both sigma formation and the formation of

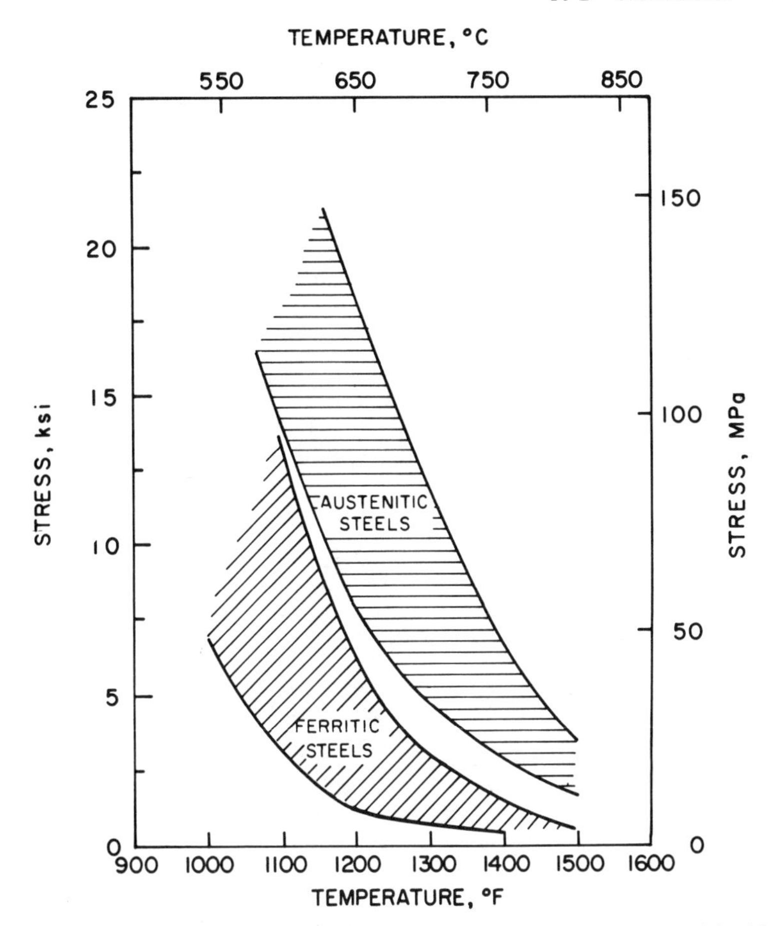

Figure 10-7. Ranges of rupture strength (rupture in 10,000 hours) for typical ferritic and austenitic stainless steels. (After Moran, Skinner, and LaQue.)

alpha prime (see Chapter 2) resulting from high temperature exposure should also be considered, particularly when impact resistance of the material on subsequent cooling is of importance.

10-2-5 Effects of Environment

The oxidation resistance of stainless steels in air is significantly decreased by the presence of water vapor. Among the earliest studies of the effect of water vapor on the oxidation resistance of stainless steels was that by Hatfield in 1927 (31), who showed that the oxidation rate of 18/8 austenitic stainless steel in air at 900°C was increased by a factor of seven by the

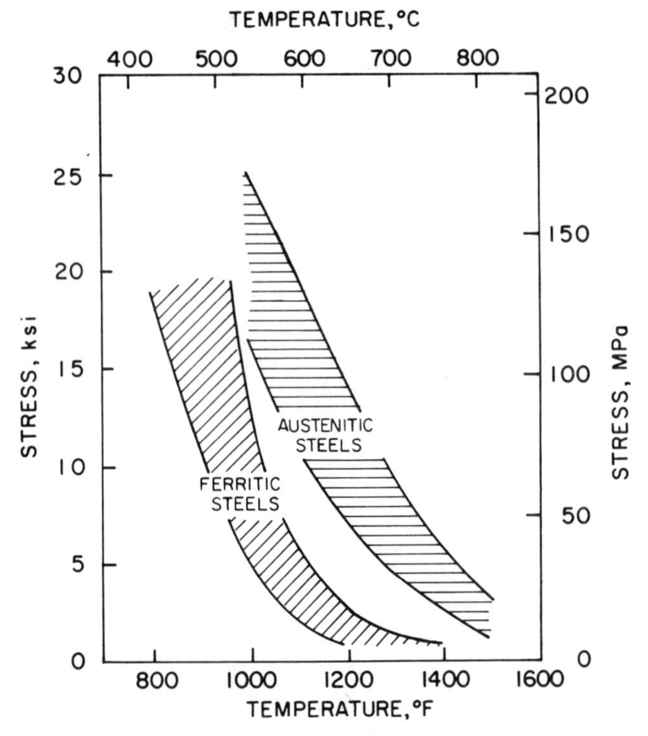

Figure 10-8. Ranges of creep strength (1% in 10,000 hours) for typical ferritic and austenitic stainless steels. (After Moran, Skinner, and LaQue.)

introduction of 5% water. There have been several studies of the effect of water vapor on the oxidation resistance of stainless steels (6, 32, 33). Studies by McCarron and Schultz (33) of the oxidation resistance of type 310 have demonstrated the importance of changes in water vapor content, as shown Figure 10-9, and have attributed the detrimental effect of water vapor to decreasing plasticity of the protective scale. In view of the detrimental effect of water vapor on oxidation resistance, Morris (2) has noted that the temperature limits for service in moist air should be adjusted downward by 38–65°C from those shown in Tables 10-2 and 10-3. For service temperature limits in high temperature steam the reader should consult the publications by Eberle et al. (32) and Cordovi et al. (34).

Interest in the oxidation behavior of stainless steels in high temperature carbon dioxide environments has been stimulated by the development of certain European gas cooled nuclear reactors. Reactors using carbon

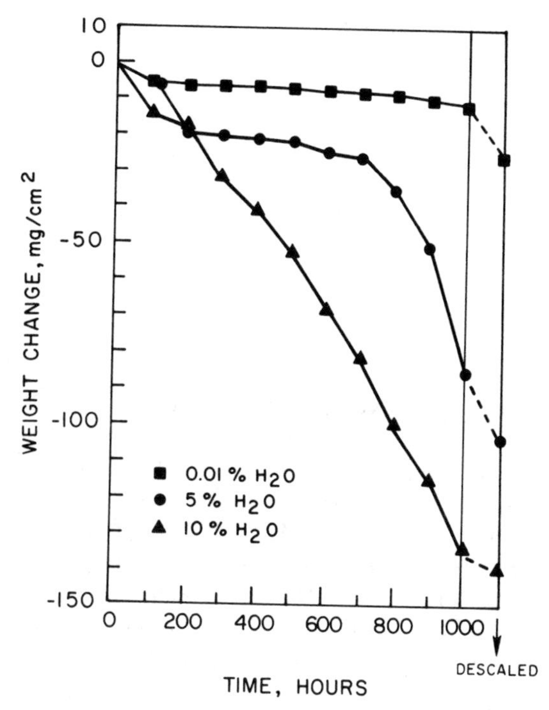

Figure 10-9. Effect of water vapor on the cyclic oxidation resistance of type 310 stainless steel at 1100°C in air. Specimens cycled every 100 hours. (After McCarron and Schultz.)

dioxide have not been developed in the U. S. For a review of the studies relating to the oxidation behavior of stainless steels in carbon dioxide at elevated temperatures the reader should consult the publication by Morris (2). This notes that for carbon dioxide the service temperatures listed in Table 10-2 (2) should be revised downward by 38–93°C for the lower chromium types 410, 430, 302, 321, 316, and 347, whereas the higher chromium types 309, 310, and 330 could be used at temperatures close to those listed for service in air, as given in Table 10-2.

10-3 SULFIDATION

10-3-1 Introduction

Sulfidation is a term used to describe attack by high temperature gases containing sulfur compounds such as sulfur dioxide, hydrogen sulfide, or sulfur vapor. The reactions can involve the formation of oxides plus sulfides in oxidizing gases such as air-sulfur dioxide mixtures, or sulfides

TABLE 10-4. MELTING POINTS OF SEVERAL METAL/METAL SULFIDE EUTECTICS

Eutectic	Melting Point, °C
$Ni-Ni_3S_2$	645
$Co-Co_4S_3$	880
Fe-FeS	985
$Al-Al_xS_x$	∿1070
Cr-CrS	∿1350

in reducing gases such as hydrogen-hydrogen sulfide mixtures. It should be noted that the terms reducing and oxidizing, when applied to the high temperature gaseous corrosion of stainless steels, relate to the behavior of chromium, and that a truly reducing condition is one under which chromium containing oxides do not form. Thus, conditions which may be reducing for other metals may in fact be oxidizing for stainless steels. Attempts to define sulfidation in terms of gas/metal reactions are further complicated by the fact that attack can sometimes be caused by molten phases such as low melting point metal/sulfide eutectics, as indicated in Table 10-4 (35).

For applications involving stainless steels, sulfidizing environments of most interest have been moist air-sulfur dioxide, hydrogen-hydrogen sulfide, sulfur vapor, combustion atmospheres generated by burning fuels containing sulfur, and atmospheres encountered in hydrodesulfurization of petroleum liquids and in coal gasification.

10-3-2 Sulfur Dioxide Environments

Although the scaling rates of high chromium stainless steels are higher in dry sulfur dioxide environments than in air, such materials are considered to have acceptable resistance to dry sulfur dioxide (2). The higher scaling rates in sulfur dioxide are attributed to the formation of chromium sulfides below the protective Cr_2O_3 scale. The exact mechanism of this subscale sulfidation remains a subject of discussion, but is thought to be associated with the regeneration of free sulfur which can react with chromium and the other alloy constituents.

The introduction of water vapor into air-sulfur dioxide atmospheres is known to accelerate attack (31), although the mechanism of acceleration is not known. It is possible that the effect of water vapor is to decrease scale plasticity, as suggested by McCarron and Schultz (33) in the case of air-water vapor environments. Recent tests by Mazzotta (36), summarized in Table 10-5, of various materials in moist air-sulfur dioxide atmospheres have confirmed the beneficial effects of increasing chromium content of stainless steels. The data of Table 10-5 also suggest that aluminum additions to nickel base alloys (i.e., Inconel alloy 601) are beneficial. However, low chromium nickel base alloys without aluminum are generally not considered for service in oxidizing atmospheres containing large amounts of sulfur (2), because of the possible formation of low melting point Ni-Ni_3S_2 eutectics (Table 10-4). The generally accepted maximum service temperatures for stainless steels in sulfur dioxide environments are shown in Table 10-6 (2).

10-3-3 Hydrogen-Hydrogen Sulfide Environments

These environments, which are encountered in catalytic reformers and desulfurizers are generally reducing to stainless steels and promote the formation of sulfide scales. Stainless steels containing less than 20% chromium (e.g., types 304, 410, and 430) tend to form iron sulfide (FeS) scales, whereas higher chromium stainless steels tend to form layered scales of various chromium and iron sulfides (37–40). It is thought that the rate controlling process in the sulfide scale growth is the transport of iron or chromium ions through the scale, which generally gives parabolic growth kinetics. Chromium and aluminum increase scaling resistance, as illustrated by the data of Table 10-7 (36). Lower chromium high nickel alloys such as HT, HU, HW, and HX are generally not used in high temperature hydrogen sulfide atmospheres because of the possibility of forming the low melting point Ni-Ni_3S_2 eutectic, given in Table 10-4, which could lead to catastrophic attack.

The corrosion rates as a function of hydrogen sulfide concentration and temperature have been defined for various martensitic and ferritic stainless steels, detailed in Table 10-8 (41), and austenitic stainless steels, shown in Figure 10-10 (42). Generally the high chromium type 310 is the favored stainless steel for high temperature hydrogen sulfide environments (2).

10-3-4 Sulfur Vapor

Stainless steels are attacked by sulfur vapor at elevated temperatures, with the higher chromium grades exhibiting greater resistance, as shown in Table 10-9. The high chromium type 310 is favored for high temperature sulfur vapor service (43).

10-3-5 Combustion Atmospheres

Most solid, liquid, or gaseous fossil fuels contain sulfur, and their combustion products often contain sulfur products. In the hot working of

TABLE 10-5. EXTENT OF ATTACK EXHIBITED BY VARIOUS STAINLESS STEELS AND HIGHER ALLOYS IN AN AIR-2% SO_2-5% H_2O ENVIRONMENT AS A FUNCTION OF TEMPERATURE DURING A 1000 HOUR ISOTHERMAL EXPOSURE

	700°C		850°C		1000°C	
Material	Weight Loss[a] (mg/cm^2)	Maximum Attack[b] (mm)	Weight Loss[a] (mg/cm^2)	Maximum Attack[b] (mm)	Weight Loss[a] (mg/cm^2)	Maximum Attack[b] (mm)
304	0	0.015	15.51	0.127	21.01	0.381
316	0.09	0.015	7.65	0.089	7.96	0.129
309	0.29	0.015	2.33	0.038	4.52	0.152
310	0.20	0.012	2.23	0.015	4.43	0.048
HK	0.31	0.010	2.23	0.017	9.05	0.175
HP	0.15	0.007	2.90	0.033	8.98	0.177
Inconel alloy 601[c]	0.26	0.010	2.35	0.066	1.89	0.246

[a]Descaled weight loss.
[b]Parameter defined in Figure 10-1.
[c]Contains 1.27%Al and 22%Cr.

TABLE 10-6. GENERALLY ACCEPTED MAXIMUM SERVICE TEMPERATURES IN SULFUR DIOXIDE FOR SELECTED STAINLESS STEELS

Material	Temperature °C
304	800
321	800
347	800
310	1050
410	700
430	800
446	1025

TABLE 10-7. CORROSION RATES OF VARIOUS MATERIALS IN A HYDROGEN-1.5% HYDROGEN SULFIDE ATMOSPHERE AT 600°C

Material	% Cr	% Al	Descaled Weight Loss, mg/cm^2	
			100 Hours	1000 Hours
304	18.3	--	37.79	126.90
310	24.0	--	32.58	97.75
INCONEL alloy 601	22.0	1.27	19.95	69.08

TABLE 10-8. CORROSION RATES OF VARIOUS STAINLESS STEELS IN HYDROGEN-HYDROGEN SULFIDE ATMOSPHERES

Test Conditions							
Temperature, °C	510	515	520	530	604	752	752
H_2 Pressure,kPa	1207	5826	3344	3344	3344	1276	1276
H_2S Conc,Vol. %	0.10	0.80	0.75	0.05	0.05	0.05	0.15
Time, hours	598	461	234	222	415	458	468
Material	Corrosion Rate, mils/yr						
405	75	--	--	--	--	13	180
410	--	220	190	300	100	--	160
430	--	39	30	60	--	16	240
446	10	15	16	42	41	76	230
300 Series	7	23	27	40	12	5	65

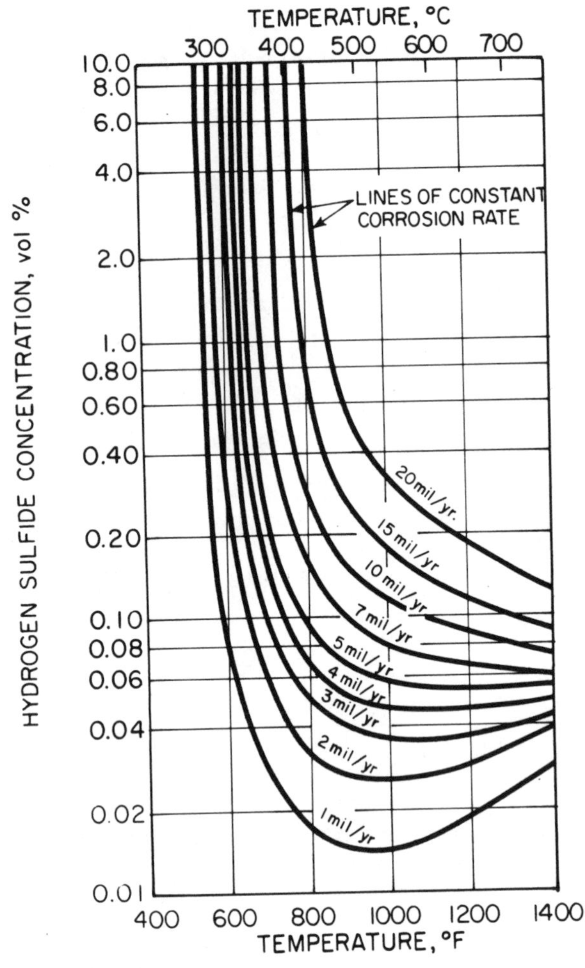

Figure 10-10. Effect of temperature and hydrogen sulfide content on the corrosion rate of austenitic stainless steels. Pressure range 1207–3447 kPa, exposure times greater than 150 hours. (After Backensto and Sjoberg.)

metals the atmospheres of ingot soaking pits and reheating furnaces are often produced by the burning of natural gas which in the U. S. has been plentiful and relatively inexpensive. Natural gas is relatively low in sulfur. Interest in using heavy oils, such as No. 2 distillate oil, which can contain 0.1–0.7% sulfur, instead of natural gas have posed questions regarding the formation of molten sulfides and hot workability of nickel containing alloys including austenitic stainless steels. In this regard, studies by Kane

TABLE 10-9. CORROSION OF STAINLESS STEELS IN SULFUR VAPOR AT 570°C

Material	Corrosion Rate* mils/yr
314	16.9
310	18.9
309	22.3
304	27.0
302B	29.8
316	31.1
321	54.8

*Corrosion rates based on 1295 hour tests.

(44) have shown that in the case of chromium containing alloys, liquid sulfides were not formed and workability was not affected when these materials were heated in high sulfur (2%) atmospheres for several hours at 1127°C. Among alloys evaluated by Kane were type 316 and A-286 stainless steels. For an analysis of the problems that can arise in the oil fired reheating of stainless steels the reader should consult the publication by Edstrom (4).

Other process and combustion gases, often referred to by the generic name of flue gas, comprise a wide range of gaseous environments that can be oxidizing or reducing. Oxidizing flue gases generally contain sulfur dioxide, carbon dioxide, nitrogen, water vapor, and oxygen. Reducing flue gases contain hydrogen sulfide, hydrogen, water vapor, carbon monoxide, carbon dioxide, and nitrogen. Because of the complexity of flue gases, field tests may be advisable for the selection of materials. Generally, increasing the chromium content of the stainless steels results in increased corrosion resistance, for example, as shown in Table 10-10 (45). For a more detailed discussion of corrosion by flue gases the reader should consult the reviews by Paul and Moran (46), Morris (2), and Jackson et al. (47).

10-4 CARBURIZATION

Carburization is the uptake of carbon by alloys and can occur when a metal is exposed to high temperature environments containing carbon bearing gases, such as carbon monoxide or hydrocarbons. In stainless steels, carbides of chromium and iron are formed when the solid solubility of carbon is exceeded, and very large volume fractions of carbide can result. Failures by carburization can be caused by the formation of these

TABLE 10-10. CORROSION RATE OF STAINLESS STEELS IN VARIOUS FLUE GASES

Material	Corrosion Rate, mils/yr*		
	Coke Oven Gas, 815°C	Coke Oven Gas, 981°C	Natural Gas, 815°C
430	91	236	12
446(26 Cr)	30	40	4
446(28 Cr)	27	14	3
302B	104	225	--
309S	37	45	3
310S	38	25	3
314	23	94	3

*Based on 3-month exposure.

carbides which may reduce ductility and toughness. Cracking can also be induced in the material below the carburized layer by the high local stresses produced by the large volume changes associated with carbide formation. Carburization occurs at elevated temperatures and is accelerated by increasing temperature.

Failures due to carburization are rare in steam hydrocarbon reforming, used for the production of hydrogen and carbon monoxide. However, in the higher temperature ethylene pyrolysis, used for the production of ethylene, carburization resistance is of prime consideration in alloy selection.

The evaluation of carburization resistance has been accomplished usually by pack carburizing or gas carburizing. In pack carburizing the test piece is surrounded by a solid carburizing compound, enclosed in a retort, and exposed to high temperatures for periods of up to 500 hours. In cyclic exposures the carburizing compound is renewed for each cycle. In gas carburizing the specimen is exposed in a high temperature furnace through which a carburizing gas mixture is allowed to flow. In our laboratory a number of gas mixtures have been used, including (*a*) dessicated 98% hydrogen-2% methane and (*b*) equilibrated hydrogen-methane-water vapor mixtures. Methods employed to measure the extent of carburization have included weight gain measurements, metallographic examination, chemical analysis of the bulk specimen, chemical analysis of thin surface layers machined from specimens, and microprobe analysis.

As in the case of oxidation and sulfidation, the alloying element that imparts the greatest resistance to carburization is chromium. Other beneficial elements are nickel, silicon, columbium, and titanium. The beneficial effects of these elements can be inferred from the comparisons

TABLE 10-11. PACK CARBURIZATION TEST*

Alloy	Nominal Composition	Silicon Content, %	Increase in Carbon Content, %**
INCOLOY alloy 800	34Ni,21Cr	0.34	0.04
330	35Ni,15Cr	0.47	0.23
330	35Ni,15Cr + Si	1.00	0.08
310	20Ni,25Cr	0.38	0.02
314	20Ni,25Cr + Si	2.25	0.03
309	12Ni,25Cr	0.25	0.12
347	8Ni,18Cr + Cb	0.74	0.57
321	8Ni,18Cr + Ti	0.49	0.59
304	8Ni,18Cr	0.39	1.40
302B	8Ni,18Cr + Si	2.54	0.22
446	28Cr	0.34	0.07
430	16Cr	0.36	1.03

*40 cycles of 25 hours at 980°C. Carburizer renewed after each cycle.
**By bulk analysis.

shown in Table 10-11 (48). Recent studies by Kane (44) of the carburization resistance of HK containing various amounts of silicon have shown the benefit of the latter to be related to the formation of a silica subscale below the carbide layer. At low oxygen partial pressures this protective silica subscale does not form. It is also believed that high carbon contents and tungsten additions enhance carburization resistance. However, no definitive studies of these elements appear to have been carried out for stainless steels. Several cast higher alloys containing chromium, nickel, silicon, and sometimes tungsten, have been developed for carburizing environments (49), as well as wrought higher alloys containing high levels of chromium, nickel and carbon, such as Incoloy alloy 802, as shown in Figure 10-11 (50).

Surface finish of castings is also believed to influence carburization resistance. However, as yet there is no mechanistic understanding of this topic, which is deserving of further study.

It should be noted that another form of metal deterioration can occur in carburizing gases, known as metal dusting. This can occur as general or localized attack, with the corrosion products being a fine powder or dust consisting of carbides, oxides, and graphite. Metal dusting is known to occur at temperatures in the range 425–815°C (2), and types 310, 316, and 302B are reportedly more resistant than types 304, 321, and 347 under similar test conditions (51). Several theories have been proposed to ex-

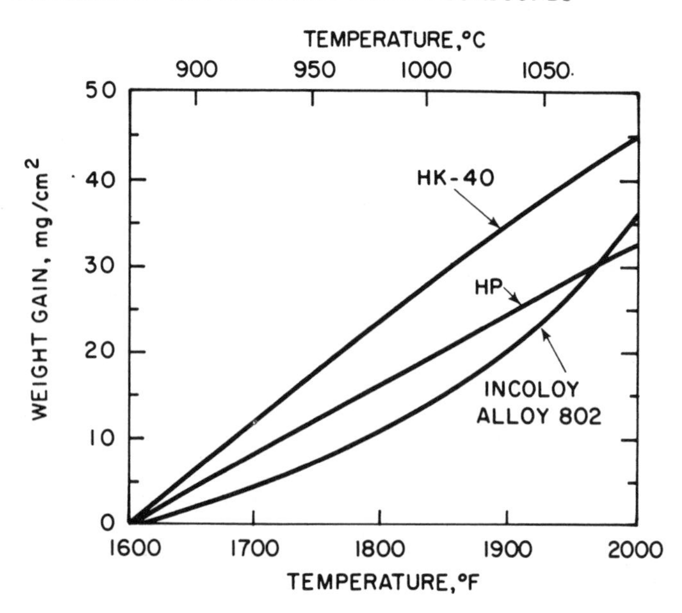

Figure 10-11. Results of 100 hour carburization tests in an atmosphere containing 2% CH_4 and 98% H_2. (After Tipton and Wenschhoff.)

plain its occurrence (52), although the phenomenon is not well understood. It is believed that metal dusting problems may be eliminated by introducing small amounts of H_2S into gas streams, although there do not appear to be any publications to describe this practice.

10-5 NITRIDING

Atomic nitrogen, which can be produced by the dissociation of ammonia at elevated temperatures, can penetrate stainless steels and form brittle nitride surface layers. Thus nitriding, like carburization, can produce a form of embrittlement. The resistance of stainless steels to nitriding depends on alloy composition and the environmental parameters of ammonia concentration and temperature. High nickel contents are beneficial, while molybdenum can be detrimental under certain conditions, as shown in Table 10-12 (53). For pure ammonia at temperatures of about 500°C, high nickel alloys are more suitable than stainless steels (53).

10-6 HALOGEN GASES

Corrosion of stainless steels and higher alloys in high temperature chlorine can be best understood by recognizing that different processes govern the corrosion rate at high and low temperatures. At high tempera-

TABLE 10-12. CORROSION RATES OF STAINLESS STEELS IN AMMONIA CONVERTER AND PLANT

Material	Nominal Ni Content, %	Corrosion Rate, mils/yr Ammonia Converter(a)	Ammonia Plant Line(b)
446	--	1.12	164.5
430	--	0.90	--
302B	10	0.73	--
304	9	0.59	99.5
316(2.23Mo)	13	0.47	>520
321	11	0.47	--
309	14	0.23	95
314	20	0.10	--
310	21	0.14	--
330(0.47Si)	34	0.06	--
330(1.00Si)	36	0.02	--

(a) 5 to 6% NH_3, 29164 h at 490 to 550°C.
(b) 99.1% NH_3, 1540 h at 500°C.

tures (e.g., 600°C) the corrosion process is governed by the formation and volatilization of the chloride scale. Thus the maximum operating temperatures are generally a function of the nickel content of the alloy, since nickel forms one of the least volatile chlorides. At lower temperatures chloride scaling rates are probably determined by diffusion through the chloride scale and the protectiveness of this scale.

For stainless steels the upper temperature limit for operation in dry chlorine and hydrogen chloride is about 320°C, as indicated in Table 10-13 (54). For higher temperature service, higher nickel alloys or pure

TABLE 10-13. CORROSION OF STAINLESS STEELS IN DRY CHLORINE AND HYDROGEN CHLORIDE AT ELEVATED TEMPERATURES

Gas	Material	Approximate Temperature, °C, for Given Corrosion Rate* 30 mils/yr	60 mils/yr	120 mils/yr
Chlorine	304	290	315	345
	316	315	345	400
Hydrogen Chloride	304	345	400	455
	316	370	370	480
	309 + Cb	345	400	455

*Corrosion rate based on short term tests of 2 to 20 hours.

TABLE 10-14. WEIGHT LOSS OF STAINLESS STEELS AND HIGHER ALLOYS ON EXPOSURE TO DRY CHLORINE-70% ARGON ATMOSPHERE AT A VELOCITY OF 16 cm/MINUTE

Material	Descaled Weight Loss After 500 Hour Exposure, mg/cm²		
	400°C	500°C	600°C
347	215	S.D.	--
321	108	S.D.	--
304	108	1110	--
310	28	370	--
HK	0.8	420	--
INCOLOY alloy 800	6	13	200-270
HASTELLOY alloy C	0.05	5	65-120
INCONEL alloy 601	0.3	3	85-200
INCONEL alloy 600	0.02	5	175-180
Nickel 201	0.2	3	47-101

S.D. = specimen destroyed.
Typical scatter of results = ±40%, unless range shown.

nickel are necessary, as outlined in Table 10-14 (55). Since at these higher temperatures volatilization of the chloride is taking place, the corrosion rate is also affected by gas flow rate, chlorine partial pressure, and specimen configuration.

The introduction of water vapor into the chlorine is known to accelerate corrosion of stainless steels at lower temperatures (54).

Dry hydrogen chloride is slightly less corrosive than dry chlorine, as shown in Table 10-13, whereas dry fluorine is more corrosive (56).

10-7 MOLTEN COMPOUNDS

10-7-1 Fuel Ash*

Fuel ash corrosion is a generic term used to describe attack by low melting sulfates and vanadates generated by burning solid and liquid fuels containing sulfur, sodium, and vanadium. The compounds may be solid while in the gas stream, but form liquid eutectics by reaction with surface oxides. In our laboratory testing for resistance to oil ash corrosion has been carried out by high temperature exposure of materials coated with 80% V_2O_5 – 20% Na_2SO_4 or 80% Na_2SO_4 – 20% V_2O_5, or by exposure to the combustion products of oils contaminated with sulfur, sodium, and

*For a more detailed treatment of this topic the reader is referred to W. T. Reid, *External Corrosion and Deposits: Boilers and Gas Turbines,* American Elsevier, N. Y., 1971.

TABLE 10-15. CORROSION RATES OF VARIOUS ALLOYS IN A CRUCIBLE TEST EMPLOYING 80% V_2O_5-20% Na_2SO_4. EXPOSURE CONDITIONS: 16 HOURS AT 750°C

Alloy	% Cr	Weight Loss, mg/cm^2
304	18.3	68.53
310	24.0	56.94
50%Ni-50%Cr	50.0	17.38

vanadium. Oil ash corrosion generally increases with temperature up to about 900°C and then decreases. Table 10-15 (36) shows that increasing the chromium content of a given alloy decreases the rate of attack. Unlike in the case of oxidation, aluminum affords no additional protection since the moten salts readily flux Al_2O_3. For a more detailed discussion of fuel ash corrosion the reader should consult references 57–59. Because of the severity of attack that can result in the presence of vanadates field tests seem advisable whenever possible.

10-7-2 Liquid Metals and Molten Salts

There have been extensive evaluations of the corrosion resistance of stainless steels in liquid metals and molten salts, in conjunction with the early development of certain prototype nuclear reactors. There is also some knowledge of the behavior of stainless steels in heat treating salt baths. For recent detailed reviews of these topics the reader should consult the reviews by Koger (58), Berry (60), and Lyon (61).

REFERENCES

1. G. C. Wood, *Corros. Sci.,* Vol. 2, p. 173, 1962.
2. L. A. Morris, in *Handbook of Stainless Steels,* D. Peckner and I. M. Bernstein, Eds., McGraw-Hill Book Co., New York, 1977, p. 17-1.
3. G. C. Wood and M. G. Hobby, *J. Iron Steel Inst.,* Vol. 203, p. 54, 1965.
4. J. O. Edstrom, *J. Iron Steel Inst.,* Vol. 185, p. 450, 1957.
5. D. Caplan and M. Cohen, *J. Met.,* Vol. 4, p. 1057, 1952.
6. D. Caplan and M. Cohen, *Corrosion,* Vol. 15, p. 141t, 1959.
7. H. J. Yearian, W. Derbyshire, and J. F. Radavich, *Corrosion,* Vol. 13, p. 597, 1957.
8. H. J. Yearian, E. C. Randell, and T. A. Longo, *Corrosion*, Vol. 12, p. 515t, 1956.
9. J. E. Croll and G. R. Wallwork, *Oxid. Met.,* Vol. 1, p. 55, 1969.
10. H. L. Eiselstein and E. N. Skinner, ASTM Special Technical Publication STP-165, American Society for Testing Materials, Philadelphia, Pa., 1954, p. 162.
11. M. G. Hobby and G. C. Wood, *Oxid. Met.,* Vol. 1, p. 23, 1969.
12. H. J. Yearian, H. E. Boren, and R. E. Warr, *Corrosion,* Vol. 12, p. 561t, 1956.
13. D. Caplan and M. Cohen, *Nature,* Vol. 205, p. 690, 1965.
14. A. Grodner, Welding Research Council Bulletin No. 31, 1956.

15. R. J. Mangone, C. J. Slunder, and A. M. Hall, *Summary Report,* ACI Project No. 32, Battelle Memorial Institute, July 15, 1958.
16. J. M. Francis, *J. Iron Steel Inst.,* Vol. 204, p. 910, 1966.
17. A. D. Brasunas and N. J. Grant, *Trans. Am. Soc. Met.,* Vol. 44, p. 1117, 1952.
18. H. Krainer, L. Wetternick, and C. Carius, *Arch. Eisenhuettenwes.,* Vol. 22, p. 103, 1951.
19. I. Kvernes, M. Oliveria, and P. Kofstad, *Corros. Sci.,* Vol. 17, p. 237, 1977.
20. H. T. Michels, "Corrosion Performance of Heat Resistant Alloys in Automobile Exhausts," *Met. Eng. Q.,* Vol. 14, No. 3, p. 23, August 1974.
21. H. L. Eiselstein and J. C. Hosier, U. S. Patent 3,729,308, 1973.
22. E. J. Felton, *J. Electrochem. Soc.,* Vol. 108, p. 490, 1961.
23. J. M. Francis and J. A. Jutson, *Corros. Sci.,* Vol. 8, p. 445, 1968.
24. J. J. deBarbadillo, The International Nickel Company, Inc., to be published.
25. S. S. Brenner, *J. Electrochem. Soc.,* Vol. 102, p. 16, 1955.
26. R. E. Keith, C. A. Siebert, and M. J. Sinnott, ASTM Special Technical Publication STP-171, American Society for Testing Materials, Philadelphia, Pa., 1955, p. 49.
27. L. Colombier and J. Hochmann, *Stainless and Heat Resisting Steels,* St. Martin's Press, New York, 1967, p. 330.
28. D. Caplan, *Corros. Sci.,* Vol. 6, p. 509, 1966.
29. J. Board, G. Holyfield, and J. Dalley, *Metallurgie,* Vol. 8, p. 23, 1968.
30. J. J. Moran, E. N. Skinner, and F. L. LaQue, *Met. Eng. Q.,* p. 1, May 1962.
31. W. H. Hatfield, *J. Iron Steel Inst.,* Vol. 115, p. 483, 1927.
32. F. Eberle, F. G. Ely, and J. A. Dillon, *Trans. ASME,* Vol. 76, p. 665, 1954.
33. R. L. McCarron and J. W. Schultz, "The Effects of Water Vapor on the Oxidation Behavior of Some Heat Resistant Alloys," *Proceedings of Symposium on High Temperature Gas-Metal Reactions in Mixed Environments,* AIME, New York, 1973, p. 360.
34. M. A. Cordovi, A. B. Wilder, J. J. B. Rutherford, and C. P. Weigel, in *Behavior of Superheater Alloys in High Temperature, High Pressure Steam,* The American Society of Mechanical Engineers, New York, 1971, p. 8.
35. K. N. Strafford, *Metal. Rev.,* Review 138, p. 153, 1969.
36. F. R. Mazzotta, The International Nickel Company, Inc., to be published.
37. M. Farber and D. M. Ehrenberg, *J. Electrochem. Soc.,* Vol. 99, p. 427, 1952.
38. A. Davin and D. Coutsouradis, *Corros.-Anticorros.,* Vol. 11, p. 347, 1963.
39. F. J. Bruns, NACE Technical Committee Report No. 57-2, 1957.
40. G. Sorell and W. B. Hoyt, NACE Technical Committee Report No. 56-7, 1956.
41. E. B. Backensto, R. D. Drew, J. E. Prior, and J. W. Sjoberg, NACE Technical Committee Report No. 58-3, 1958.
42. E. B. Backensto and J. W. Sjoberg, *Corrosion,* Vol. 15, p. 125t, 1959.
43. J. R. West, *Chem. Eng.,* Vol. 58, p. 276, 1951.
44. R. H. Kane, The International Nickel Company, Inc., to be published.
45. W. F. White, *Mater. Prot.,* Vol. 2, p. 47, 1963.
46. G. T. Paul and J. J. Moran, in *Corrosion Resistance of Metals and Alloys,* Reinhold Publishing Corporation, New York, 1963, p. 375.
47. J. H. Jackson, C. J. Slunder, O. E. Harder, and J. T. Gow, *Trans. ASME,* p. 1021, August 1953.
48. J. F. Mason, J. J. Moran, and E. N. Skinner, *Corrosion,* Vol. 16, p. 593t, 1960.
49. H. S. Avery (p. 61), and A. R. Ward (p. 189), in *Materials Technology in Steam Reforming Processes,* C. Edeleanu, Ed., Pergamon Press, New York, 1966.
50. D. G. Tipton and D. E. Wenschhoff, *Carburization Testing of Some Ethylene Furnace Alloys,* Paper No. 9, presented at NACE Corrosion/76, Houston, Tex., March 1976.

51. F. A. Prange, *Corrosion,* Vol. 15, p. 619t, 1959.
52. I. Koszman, "Factors Affecting Catastrophic Carburization (Metal Dusting)," *Proceedings of Symposium on High Temperature Gas-Metal Reactions in Mixed Environments,* AIME, New York, 1973, p. 155.
53. J. J. Moran, J. R. Mihalisin, and E. N. Skinner, *Corrosion,* Vol. 17, p. 191t, 1961.
54. M. H. Brown, W. B. DeLong, and J. R. Auld, *Ind. Eng. Chem.,* Vol. 39, p. 839, 1947.
55. R. H. Kane and J. E. Chart, The International Nickel Company, Inc., to be published.
56. W. R. Meyers and W. B. DeLong, *Chem. Eng. Prog.,* Vol. 44, p. 359, 1948.
57. *High Chromium, High Nickel Alloys for High Temperature Corrosion Resistance,* The International Nickel Company, Inc., New York, 1968.
58. J. W. Koger, in *Handbook of Stainless Steels,* D. Peckner and I. M. Bernstein, Eds., McGraw-Hill Book Co., New York, 1977, p. 18-1.
59. G. L. Swales, *Australas. Corros. Eng.,* Vol. 7, p. 3, September 1963.
60. W. E. Berry, *Corrosion in Nuclear Applications,* John Wiley & Sons, Inc., New York, 1971.
61. R. N. Lyon, *Liquid Metals Handbook,* U. S. Government Printing Office, Washington, D. C., 1952.

APPENDIX

1. TRADEMARKS

Trademarks	*Producer*
AL	Allegheny Ludlum Steel Corp.
Allegheny	Allegheny Ludlum Steel Corp.
Almar	Allegheny Ludlum Steel Corp.
AM	Allegheny Ludlum Steel Corp.
Carpenter	Carpenter Technology Corp.
Chlorimet	Duriron Co.
Croloy	Babcock & Wilcox Co.
Crucible	Colt Industries, Inc.
Custom	Carpenter Technology Corp.
E-Brite	Allegheny Ludlum Steel Corp.
Haynes	Cabot Corp.
Hastelloy	Cabot Corp.
HNM	Colt Industries, Inc.
Illium	Stainless Foundry & Eng. Co.
Incoloy	Inco family of companies
Inconel	Inco family of companies
Jessop	Jessop Steel
Monel	Inco family of companies
Multiphase (MP)	Standard Pressed Steel Co.
Ni-Span-C	Inco family of companies
Nitronic	Armco Steel Corp.
René 41	Teledyne Allvac
Sandvik	Sandvikens Jernverks Aktiebolag
Stellite	Deloro Stellite Ltd.
Tenelon	United States Steel Corp.
Uddeholm (UHB)	Uddeholms Aktiebolag
Uniloy	Universal Cyclops Steel Corp.

Unitemp	Universal Cyclops Steel Corp.
Waspaloy	United Technologies Corp.
PH15-7Mo	Armco Steel Corp.
17-7PH	Armco Steel Corp.
17-4PH	Armco Steel Corp.
PH13-8Mo	Armco Steel Corp.
18SR	Armco Steel Corp.
MF-2	Allegheny Ludlum Steel Corp.
OR-1	Allegheny Ludlum Steel Corp.
A-286	Allegheny Ludlum Steel Corp.

2. TYPICAL PROPERTIES OF WROUGHT STAINLESS STEELS

Typical Properties of Wrought Stainless Steels

To convert to metric: 1000 psi = approx. 6.9 MPa.
C° = (°F—32) ÷ 1.8
1 Ft.-Lb. = approx. 1.4 joules

AISI Type	Typical Composition, % (a) Max. (if not designated otherwise)	Form (b)	Mechanical Properties of Annealed Material at Room Temperature: Tensile Strength, 1000 Psi	Yield Strength (0.2% Offset), 1000 Psi	Elongation in 2 in., %	Hardness	Izod Impact Strength Ft-Lb	Mechanical Properties at Elevated Temperatures: Creep Strength, Load for 1% Elongation in 10,000 Hr, 1000 Psi: 1000 F	1100 F	1200 F	1300 F	1500 F	Scaling Temperature: Max Continuous Service in Air, F	Max Intermittent Service in Air, F	Thermal Treatment: Initial Forging Temperature, F	Annealing Temperature, F (d)	Stress-Relief Annealing Temperature, F	Melting Range, F
Austenitic (c)																		
201	16-18 Cr, 3.5-5.5 Ni, 0.15 C, 5.5-7.5 Mn, 1.0 Si, 0.060 P, 0.030 S, 0.25 N	Sheets Strips Tubing	115 115 115	55 55 55	55 55 55	Rb 90 Rb 90 Rb 90	110-120	—	—	—	—	—	1550	1450	2100-2250	1850-2050	—	—
301	16-18 Cr, 6-8 Ni, 0.15 C, 2.0 Mn, 1.0 Si, 0.045 P, 0.030 S	Plates Sheets Strips Tubing	105 110 110 105	40 40 40 40	55 60 60 50	Bhn 165 Rb 85 Rb 85 Rb 95	100	19	12.5	8	4.5	1.8	1650	1500	2100-2300	1850-2050	400-750	2250-2590
302	17-19 Cr, 8-10 Ni, 0.15 C, 2.0 Mn, 1.0 Si, 0.045 P, 0.030 S	Bars Plates Sheets Strips Tubing Wire	85 90 90 90 85 90	35 35 40 40 35 35	60 60 50 50 50 60	Bhn 150 Rb 80 Rb 85 Rb 85 Rb 85 Rb 83	110	20	12.5	7.5	4.3	1.5	1650	1500	2100-2300	1850-2050	400-750	2550-2590
303	17-19 Cr, 8-10 Ni, 0.15 C, 2.0 Mn, 1.0 Si, 0.20 P, 0.15 S min, 0.60 Mo (optional)	Bars Tubing Wire	90 80 90	35 38 35	50 53 50	Bhn 160 Rb 76	85	16.5	11.5	6.5	3.5	0.7	1650	1400	2100-2350	1850-2050	400-750	2550-2590
303Se	17-19 Cr, 8-10 Ni, 0.15 C, 2.0 Mn, 1.0 Si, 0.20 P, 0.06 S, 0.15 Se min																	
304	18-20 Cr, 8-10.50 Ni, 0.08 C, 2.0 Mn, 1.0 Si, 0.045 P, 0.030 S	Bars Plates Sheets Strips Tubing Wire	85 82 84 84 85 90	35 35 42 42 35 35	60 60 55 55 50 60	Bhn 149 Bhn 149 Rb 80 Rb 80 Rb 80 Rb 83	110	20	12	7.5	4	1.5	1650	1550	2100-2300	1850-2050	400-750	2550-2650
304L	18-20 Cr, 8-12 Ni, 0.03 C, 2.0 Mn, 1.0 Si, 0.045 P, 0.030 S	Plates Sheets Strips Tubing	79 81 81 78	33 39 39 34	60 55 55 55	Bhn 143 Rb 79 Rb 79 Rb 75												
305	17-19 Cr, 10.50-13 Ni, 0.12 C, 2.0 Mn, 1.0 Si, 0.045 P, 0.030 S	Plates Sheets Strips Tubing Wire	85 85 85 80 85	35 38 38 36 74	55 50 50 56 60	— Rb 80 Rb 80 Rb 80 Rb 77	110	19	12.5	8	4.5	2	1650	—	2100-2300	1850-2050	—	2550-2650
308	19-21 Cr, 10-12 Ni, 0.08 C, 2.0 Mn, 1.0 Si, 0.045 P, 0.030 S	Bars Plates Sheets Strips Tubing Wire	85 85 85 85 85 95†	30 30 35 35 35 60†	55 55 50 50 50 50†	Rb 80 Bhn 150 Rb 80 Rb 80 Rb 80	110	—	—	—	—	—	1700	1550	2100-2300	1850-2050	—	2550-2590

309	22-24 Cr, 12-15 Ni, 0.20 C, 2.0 Mn, 1.0 Si, 0.045 P, 0.030 S	Bars Plates Sheets Strips Tubing Wire	95 95 90 90 90 105†	40 40 45 45 45 70†	45 45 45 45 45 35†	Rb 83 Bhn 170 Rb 85 Rb 85 Rb 85 Rb 98†	110	16.5	12.5	10	6	3	1950	1850	2050-2250	1900-2050	—	2550-2650
309S	22-24 Cr, 12-15 Ni, 0.08 C, 2.0 Mn, 1.0 Si, 0.045 P, 0.030 S																	
310	24-26 Cr, 19-22 Ni, 0.25 C, 2.0 Mn, 1.5 Si, 0.045 P, 0.030 S	Bars Plates Sheets Strips *Tubing Wire	95 95 95 95 95 105†	45 45 45 45 45 75†	50 50 45 45 45 30†	Rb 89 Bhn 170 Rb 85 Rb 85 Rb 85 Rb 98†	110	33	23	15	10	3	2050	1900	2000-2250	1900-2100	400-750	2550-2650
310S	24-26 Cr, 19-22 Ni, 0.08 C, 2.0 Mn, 1.5 Si, 0.045 P, 0.030 S																	
316	16-18 Cr, 10-14 Ni, 0.08 C, 2.0 Mn, 1.0 Si, 0.045 P, 0.030 S, 2.0-3.0 Mo	Bars Plates Sheets Strips Tubing Wire	80 82 84 84 85 80	30 36 42 42 35 30	60 55 50 50 50 60	Rb 78 Bhn 149 Rb 79 Rb 79 Rb 85 Rb 78	110	25	17.4	11.6	7.5	2.4	1650	1550	2100-2300	1850-2050	400-750	2500-2550
316L	16-18 Cr, 10-14 Ni, 0.03 C, 2.0 Mn, 1.0 Si, 0.045 P, 0.030 S, 2.0-3.0 Mo	Plates Sheets Strips Tubing	81 81 81 80	34 42 42 35	55 50 50 55	Bhn 146 Rb 79 Rb 79 Rb 78												
321	17-19 Cr, 9-12 Ni, 0.08 C, 2.0 Mn, 1.0 Si, 0.045 P, 0.030 S (Ti, 5×C min)	Bars Plates Sheets Strips Tubing Wire	85 85 90 90 85 95†	35 30 35 35 35 65†	55 55 45 45 50 40†	Bhn 150 Bhn 160 Rb 80 Rb 80 Rb 80 Rb89†	110	18	17	9	5	1.5	1650	1550	2100-2300	1750-2050	400-750(e)	2550-2600
Ferritic (c)																		
430	16-18 Cr, 0.12 C, 1.0 Mn, 1.0 Si, 0.040 P, 0.030 S	Bars Plates Sheets Strips Tubing Wire	75 75 75 75 75 70	45 40 50 50 40 40	30 30 25 25 25 35	Bhn 155 Bhn 160 Rb 85 Rb 85 Rb 80 Rb 82	35	8.5	4.7	2.6	1.4	—	1550	1650	1900-2050	Low anneal 1400-1500	—	2600-2750
430F	16-18 Cr, 0.12 C, 1.25 Mn, 1.0 Si, 0.060 P, 0.15 S min, 0.60 Mo (optional)	Bars Wire	80 95†	55 85†	25 10†	Bhn 170 Rb 92†	5-50	8.5	4.6	1.9	1.3	—	1500	1600	1950-2100	Low anneal 1250-1400	—	2600-2750
430FSe	16-18 Cr, 0.12 C, 1.25 Mn, 1.0 Si, 0.060 P, 0.060 S, 0.15 Se min																	
Martensitic (c)																		
410	11.5-13.5 Cr, 0.15 C, 1.0 Mn, 1.0 Si, 0.040 P, 0.030 S	Bars Plates Sheets Strips Tubing Wire	75 70 70 70 75 75	40 35 45 45 40 40	35 30 25 25 30 30	Rb 82 Bhn 150 Rb 80 Rb 80 Rb 82 Rb 82	85	11.5	4.3	2	1.5	—	1300	1450	2000-2200(f)	1500-1650(g) 1200-1400(h)	H1700-1850(d) T 400-1400(i)	2700-2790

(a) Single values are maximums, except as noted; (b) Forms listed are only those for which mechanical properties are given. Most types are available in many forms; (c) Austenitic, hardenable by cold working; not hardenable by heat treatment. Ferritic, not hardenable by heat treatment or cold working. Martensitic, hardenable by heat treatment; (d) Followed by rapid cooling. H is hardening temperature; T is tempering; (e) Stabilizing temperature, 1550 to 1650 F; (f) Retarded cool; (g) Full anneal, followed by slow cooling; (h) Low anneal; (i) Tempering within the range of 800 to 1100 F is not recommended because of resulting low and erratic impact properties and reduced corrosion resistance. Time at temperature and temperatures may vary depending on part size; (j) Retarded cool and anneal.

*Composition for Type 310 tubing varies slightly from AISI values. For standard compositions, refer to ASTM A213.

†Soft temper.

INDEX